LEAF PROTEIN
AND OTHER ASPECTS OF FODDER FRACTIONATION

LEAF PROTEIN
AND OTHER ASPECTS OF
FODDER FRACTIONATION

N. W. PIRIE

CAMBRIDGE UNIVERSITY PRESS

CAMBRIDGE

LONDON · NEW YORK · MELBOURNE

Published by the Syndics of the Cambridge University Press
The Pitt Building, Trumpington Street, Cambridge CB2 1RP
Bentley House, 200 Euston Road, London NW1 2DB
32 East 57th Street, New York, NY 10022, USA
296 Beaconsfield Parade, Middle Park, Melbourne 3206, Australia

© Cambridge University Press 1978

First published 1978

Printed in Great Britain by
Hazell Watson & Viney Ltd, Aylesbury, Bucks

Library of Congress Cataloguing in Publication Data
Pirie, Norman W., 1907–
Leaf protein and other aspects of fodder fractionation.
Includes bibliographical references and index.
1. Plant proteins. 2. Leaves – Composition.
3. Forage plants – Composition. 1. Title.
TP453.P7P53   641.1'2   77–87387.
ISBN 0 521 21920 5

Contents

	Preface	vii
	List of abbreviations	xi
1	Historical and anatomical background	1
2	Prelude to production	8
3	Choice of crops and yields of extractable protein from them	15
4	Separation, purification, composition and fractionation	46
5	Preservation, storage and modification	68
6	Digestibility *in vitro* and nutritive value in animals	75
7	Human trials and experiments	98
8	The value of the extracted fibre and the 'whey'	112
9	The role of fodder fractionation in practice	124
	Appendix: Hilaire Marin Rouelle (1718–79)	138
	References	142
	Author index	170
	Subject index	177

Preface

The suggestion that it would often be advantageous to fractionate leafy crops was for many years received with scepticism, or even hostility. It is now gaining acceptance, partly because the recent increase in the cost of oil has focussed attention on the need to waste less fuel, and partly for the familiar reason that people who were exposed to the suggestion when they were young are now reaching influential positions. Though welcome, this acceptance encourages the mistaken assumption that we now have adequate knowledge about what should be grown for fractionation, how it should be fractionated, and how the products should be used. This is far from the case. The amount of research that has been done is trivial compared to the amount that has been, and is being, done on projects of comparable complexity and smaller potential yield, such as the cultivation of microorganisms or the processing of fish. Some of the more obvious topics on which much more research is needed are outlined in this book. It is well to remember that haymaking and ploughing are ancient arts on which useful research is still being done.

The original objective of most work on fodder fractionation was the extraction, from leaves, of protein that could be used as human food. During the last decade, the assertion has often been made that there is no need for protein concentrates in human diets. This is obviously untrue in regions where the staples are bananas, cassava, yams and similar foods containing little protein. It is probably also untrue even when wheat, rice or potatoes are the staples. This is not the place to discuss the vexed question of human protein requirement: it has been discussed elsewhere (e.g. Pirie, 1976a; Scrimshaw, 1976). There are signs that orthodox opinion is now moving towards a more generous assessment of protein requirement. And the future is uncertain. As Allen (1976) puts it: 'Today we are far more unsure concerning the prospects for the world food situation than at any time since the early

1950s.' No reasonable resource should therefore be left unstudied. Furthermore, if protein concentrates were not needed, there would be no need for animal husbandry, fishing, or separating protein from oilseeds such as soy.

Terminology is important in every subject. Agreement about it is convenient, but it is more important to ensure that it is neither ambiguous nor misleading. There is no ambiguity about the equivalent words 'juice' and 'extract'. The coagulum separated from leaf juice, but not further fractionated, is called leaf protein (LP) in this book. It is often called leaf protein concentrate (LPC) by others. LP is a mixture of many proteins; the intrusion of the word *concentrate* suggests that a product is being referred to which has a greater content of true protein than some parent substance that would be called leaf protein. It can be argued that it should be called leaf lipoprotein because it contains 30 to 40 % of non-protein material – most of it lipid. That would imply that most of the lipid was attached to, rather than merely mixed with, most of the protein. This is probably not the case. The matter is of no great importance and LP is shorter. It is of more importance that the residue from which some protein has been extracted should not get a misleading name. Here, it is called fibre – which is brief. It could be called extracted residue. The most misleading name for it is pressed crop: that perpetuates the widespread illusion that pressing is the important feature of fractionation. Here, juice from which protein has been removed is called 'whey'; that is brief and the metaphor is obvious. It is sometimes longwindedly, but not misleadingly, called brown juice or deproteinised juice.

In some publications it is not clear whether the yields given are for moist press-cake of LP, dry LP, or the true protein component of the LP. Here, wherever the contrary is not stated, the yields given are for 100 % protein, and they are usually calculated by multiplying a measurement of nitrogen by six.

Advice and unpublished information from many friends who have worked with me on problems connected with fodder fractionation have been extremely valuable. My thanks are therefore due to: R. M. Allison, D. B. Arkcoll, F. C. Bawden, R. A. Buchanan, J. B. Butler, M. Byers, E. M. Crook, M. N. G. Davys, R. P. Devadas, G. N. Festenstein, E. M. Holden, R. N. Joshi, G. Kamalanathan, S. Matai,

O. L. Oke, F. H. Shah, N. Singh, M. A. Stahmann, R. L. M. Synge and M. V. Tracey. Obviously they do, or did, not necessarily agree with all the conclusions that have been drawn, or the opinions expressed.

List of abbreviations

ARC	Agricultural Research Council
cm	centimetre
DGLV	dark green leafy vegetable
DM	dry matter
FAO	Food and Agriculture Organisation of the United Nations
g	gram
ha	hectare
IBP	International Biological Programme (this has now ended)
ICI	Imperial Chemical Industries
J	joule$=1$ watt second$=0.239$ calories (small)
kg	kilogram
km	kilometre
kPa	kilo Pascal$=1/98$ kg cm^{-2}
kW h	kilowatt hour
LP	leaf protein (unfractionated)
m	metre
ml	millilitre
mm	millimetre
MJ	mega joule$=1$ million joules$=239$ kilocalories
M£	1 million pounds sterling
μm	millionth of a metre
NEDO	National Economic Development Office
t	tonne
TCA	trichloroacetic acid
UNESCO	United Nations Educational, Scientific and Cultural Organisation
USDA	United States Department of Agriculture
WHO	World Health Organisation

1
Historical and anatomical background

It is seldom possible to date precisely the beginning of a new line of research. Preliminary vague hints, coming from folk medicine, primitive technology or some such source, are as a rule slowly integrated into a new fabric. Leaf protein is different. The date of the first publication on it is 1773. The chemical category now known as 'protein' was not recognised at that time: the word itself was not coined until 1838 when Berzelius used it in a letter to Mulder. The main criterion used in 1773 was the one that Beccari had used in 1728 (not published till 1747) when he recognised the similarities between gluten from wheat and various animal products – they stank similarly when putrid or when heated. The 'smell of burnt feathers' retained a place in text books, as a means of protein recognition, for many years. Berthollet demonstrated the presence of nitrogen in this group of substances in 1785, the xanthoproteic colour test was described by Fourcroy and Vauquelin in 1800, and Fourcroy stressed the importance of nitrogen in the nutrition of plants and animals in about 1806. These observations were reasonably factual. Then, as was his way, Liebig confused the issue by using his imagination rather than his undoubted experimental skill. He decided that there were only four proteins and asserted that the curds separating from boiled vegetable juices were indistinguishable from those separating from such animal fluids as blood and egg white. As Berzelius remarked in 1843: 'This easy kind of physiological chemistry is created at the writing desk, and is the more dangerous, the more genius goes into its execution.' The individuality of proteins was not established till many years later.

The brothers Rouelle, though very unlike in temperament and appearance, are often confused. Their portraits have even been misassigned. Guillaume François (1703–70) was an enthusiastic and excitable teacher. His manner of teaching was described as '... *souvent incorrecte et familière, mais toujours animée et pittoresque, ...*'

and he enlivened proceedings on some occasions by partly undressing, and by explosions. Even when young he was megalomaniac, accusing many contemporaries of plagiarism, using such phrases as '*Ecoutez-moi! car je suis le seul qui puisse vous démontrer ses verités*'; in 1756 he claimed to have a secret weapon with which he could destroy London and the British navy. Patriotic feeling was so strong that he refused a tempting offer from a London publisher; his lectures remained unpublished. Lavoisier was among his distinguished pupils, and it is after him that the rue Rouelle in Paris is named. Because of increasing eccentricity he resigned his professorship at the Jardin du Roi (now Jardin des Plantes) in 1768. At his request, his brother, Hilaire Marin (1718–79), succeeded him – but only as demonstrator.

By contrast, H. M. Rouelle was neat and tactful; an experimenter rather than a forceful teacher. He is less often mentioned in the dictionaries in spite of a distinguished research record. He found potassium in 'cream of tartar' and went on to isolate tartaric acid from unfermented grape juice and to make tartarates of several metals. He found formic acid in ants, and made urea, or possibly the hydrate of urea and sodium chloride, from urine. This was not a novelty: Boerhaave probably made it in 1729. He confirmed the presence of iron in blood. This hardly needed confirmation, for the removal of rust stains from blood-soiled clothing must have been a familiar problem. He extended his brother's studies on various oils and resins. At that time, starch and fat were thought to be the nutritionally important components of foodstuffs. Rouelle, having begun to recognise proteins as a chemical category, stressed their nutritional importance. Clearly, Rouelle helped to establish the science of biochemistry, but he had not quite shaken off the older outlook. For example, he did not completely reject the possibility of alchemical transformations, and argued that, although we could not make an animal or a plant, it did not follow that a metal, without structure or parts and therefore presumably lifeless, could not be made. The argument that the characteristic feature of an organism was its heterogeneity had already been used by Jean Rey in 1630 (cf. Pirie, 1964*a*).

Rouelle (1773) published two papers on leaf protein (LP). The second contains so much information that it can be taken as the origin of our subject. Its title, '*Sur les Fécules ou parties vertes des*

Plantes, & sur la matière glutineuse ou végéto-animale', shows his recognition that plant and animal products are similar. He pounded leaves of several species in a marble mortar, pressed out the juice and heated it. A green coagulum, which he could decolourise by washing with alcohol, separated at about the temperature at which he could no longer keep his finger in the hot juice.* He filtered off that coagulum and got a pale coagulum on further heating. Although this paper is now often referred to, and since about 1952 the correct initials are usually given to its author, it seems to have been seldom read. Patent Office inspectors are particularly remiss and allow patents that cover points clearly established in 1773! I therefore give a free translation of the paper in an appendix to this book. Rouelle's work was extended by Vauquelin and Fourcroy (1789) a few years later.

In spite of the defects and tediousness of contemporary methods for measuring nitrogen, Boussingault concluded in 1839 that atmospheric nitrogen (N_2 or dinitrogen) and food nitrogen were not interchanged during animal metabolism. The generalisation that dinitrogen is little, if at all, used by animals, or made by them from their food, is still accepted by most scientists though there are a few puzzling apparent exceptions. Mulder attached particular importance to plant proteins as the source from which animals derived their protein and other nitrogenous substances. An animal, as Johnson phrased it in 1867, 'moulds over these vegetable principles into the fibrine, albumin and casein of its muscle and other tissues, of its blood, milk and other secretions'. As a step towards assessing their value as fodder, forages and other animal feeding stuffs were analysed in many laboratories – notably at Rothamsted by Lawes and Gilbert.

During the nineteenth century there are only casual references to the extraction of protein from leaves; extraction from seeds was more actively pursued. Extraction from leaves had probably been suggested by someone because Lawes (1885) remarked: 'It might be possible by some chemical process to produce from grass a nutritious substance which a man could use as food, but the food so extracted would be

---

* Fahrenheit made thermometers and developed the idea of a temperature scale in 1742. Martel described the Centigrade scale (sometimes absurdly called the Celsius scale) in 1742. Meteorologists adopted thermometers quickly – but chemists did not. They are, for example, not mentioned in Macquer's *Dictionnaire de chymie* (1766).

far more costly than as it existed in the grass, and no one would think of preparing such a food for oxen or sheep, as their machinery is quite competent to separate the nutritious from the indigestible portion of the food.' Lawes' comment on ruminants is justified: it is fortunate that his comment on the extraction of human food was not widely known when work on LP extraction started at Rothamsted!

Winterstein (1901) extracted protein from dried, ground leaves with dilute alkali. Sustained work started 20 years later when Osborne temporarily forsook his studies on the seed proteins. Osborne & Wakeman (1920) and Osborne, Wakeman & Leavenworth (1921) made preparations from spinach (*Spinacia oleracea*) and lucerne (*Medicago sativa*) by pulping the fresh leaf, removing the coarser particles by centrifuging or filtering, and coagulating with alcohol.

Osborne (1924) was well aware of the importance of this work, but he was distressed by the properties of the material. As he put it: 'Our present meagre knowledge of the protein constituents of living plants is chiefly due to the difficulties encountered in separating the contents of the cells from the enveloping walls. Attempts to grind the fresh leaf and extract the contents of the cells with water result in mixtures that cannot be filtered clear, and consequently appear to present no opportunity to obtain the protein in a state fit for chemical examination.' As Vickery (1956) put it: '... when Chibnall came to the laboratory in 1922 for a two-year period, Osborne was happy to turn this difficult problem over to him ...'. Chibnall had already worked with extracts from cabbage (*Brassica oleracea*) and runner bean (*Phaseolus vulgaris*). One chapter in his 1939 book, *Protein metabolism in the plant*, describes his work.

The work of Chibnall and Osborne established LP as a reasonable material for biochemical investigation. A steady flow of papers on the techniques of extraction in the laboratory soon started, e.g. Davies (1926); Lugg (1932, 1939); Kiesel *et al.* (1934); Yemm (1937); Foreman (1938); Crook (1946); Crook & Holden (1948); Holden & Tracey (1948); Bryant & Fowden (1959); and Festenstein (1961). Because leaves from different species, and of differing age and nutritional status, were studied the conclusions come to in these papers are not identical. However, the general conclusion was that the younger the leaf and the greater its protein and water content, the greater the percentage extraction of protein. Also, thorough sub-division or rubbing, and the maintenance of alkaline conditions during the extraction, are advantageous.

During the past quarter century the total amount of biochemical research has increased enormously. The amount of research on plants has not increased correspondingly, and that research tends to be concerned with alkaloids, pigments and the components of seeds and tubers. Nevertheless, much more work has been done on proteins in leaves than it would be useful to describe in any detail in a book primarily concerned with the practical problems arising when an edible protein is being extracted in bulk. Early academic work on LPs is described in several reviews, e.g. Vickery (1945), Pirie (1955, 1959a), and various specialised aspects of the subject are covered regularly in the review journals. It may, nevertheless, be useful to give a brief survey of the subject so as to explain the reasons for some of the methods and precautions adopted during the production of LP.

Even casual study of pieces of leaf under the microscope shows that the green colour is concentrated in chloroplasts: the colour of leaf extracts shows that chloroplasts, or their fragments, are being extracted. Methods for separating them from the other types of protein will be discussed later (p. 62). Here all that need be said is that there is much research on their isolation in physiologically active forms. Those from some species are extremely fragile, while from others they are robust enough to withstand the disintegration of the leaf during digestion by *Clostridium roseum* (White *et al.*, 1948). The durability of chloroplasts depends on the environment into which they are put. Extreme examples are the prolonged photosynthetic activity of chloroplasts from higher plants introduced artificially into mammalian cells (Nass, 1969), and the frequent survival of algal chloroplasts in molluscs (Taylor, 1968; Trench, Boyle & Smith, 1973; Hinde & Smith, 1975). When juice is pressed out from leaf pulp, the extent to which chloroplasts are extracted depends on their integrity and on the compaction of the leaf fibre. Typical chloroplasts are 5 μm wide and 2 μm thick. It is therefore advantageous to choose species with fragile chloroplasts, to keep the layer of pulp thin, and to express gently as much juice as possible so as to avoid pressing the fibre into a relatively impermeable mat. These points are important because much, probably most, of the protein is in the chloroplasts initially.

The smaller cellular components, such as mitochondria and ribosomes, also contain protein and they are less likely to be held back by compacted fibre. Furthermore, they disintegrate readily. This is doubly advantageous; it prevents loss and it diminishes the amount of nucleic acid that separates along with the LP. Extracts from young leaves may contain 2 or 3 g of ribosomes per litre, and these contain 20 to 30 % of RNA (Pirie, 1950). Provided the diet does not contain much RNA from other sources, the quantity that would be eaten in LP would be harmless, but, especially for people with a tendency to gout, all sources of nucleic acid must be watched. Leaf extracts contain ribonuclease so, as will be explained later (pp. 17, 49), nucleic acid is easily degraded.

Many enzymes are present in leaf extracts; their molecules are mostly too small to be retained by a packed mat of fibre to a serious extent. The dominant component is ribulose-1,5-diphosphate carboxylase (often called fraction 1 protein) which is an important component of the photosynthetic mechanism. Preparations of this enzyme from different species have similar amino acid compositions; that similarity, together with the large number of different enzymes making up LP, probably explains the constancy of its composition (p. 58). Problems arise in handling leaf extracts because of the presence of enzymes that catalyse the oxidation of phenolic substances and produce tanning agents that coagulate protein and diminish its digestibility.

However thoroughly leaves are ground, some nitrogen remains attached to the fibre. Part of this is probably protein that was initially soluble but gets attached to the fibre as a result of the stresses (perhaps momentary local heating) during grinding (Bawden & Pirie, 1944; Crook, 1946). The fibre residue is always bright green, and therefore has some chloroplasts entangled in it. Holden & Tracey (1950) measured the ratio of chlorophyll to nitrogen in preparations of chloroplasts and extracted fibre, and concluded that 80 % of the nitrogen could be accounted for by entangled chloroplasts. They suggested that the change in texture that takes place when fibre is incubated with proteolytic enzymes could be the result of removal of nonchloroplast protein from cell walls. This suggestion gains strong support from the recognition of hydroxy proline (a characteristic component of cell walls elsewhere) in the fibrous residue (Clarke & Ellinger, 1967; Jennings & Watt, 1967). Laird, Mbadiwe & Synge (1976), working with young lucerne leaves, found that the amount of nitrogen that remained insoluble even after extraction with a mixture of phenol, acetic acid and water, was more than one-third of the amount that was present in readily extractable protein.

Those who are interested in extracting protein in bulk are usually satisfied if 50 to 70 % of the protein nitrogen is being extracted, and choose species from the young leaves of which that degree of extraction is possible. The precise nature of the unextracted nitrogen, or nitrogen extracted with difficulty, is, however, of great interest to those wishing to use the residue as cattle fodder.

2
Prelude to production

Lawes (1885) suggested the possibility of making human food from inedible leaves: Ereky (1927) took out the first patent on the idea. This was the forerunner of patents by many others for methods for making food for nonruminants from material that could otherwise be used as ruminant food only. Ereky, who was Minister of Development during the Horthy dictatorship in Hungary, tried to get an article on feeding ducks and pigs on his product published in a British journal in 1924 and 1925. He was secretive about his process, because of the impending patent, and the article seems not to have been published. Three-quarters of his patent specification is devoted to confused and largely erroneous statements about the changes undergone by leaves during autolysis and drying. The machine was designed to cut, rather than crush, forage between knives set on opposed faces of a pair of truncated cones. The forage was carried round (as in the Hollanders used in paper making) in a stream of water which was recycled after separating the LP by an unspecified method. In spite of his official position, but not unexpectedly considering the inadequacy of his machine, he was unable to make enough LP for the continued feeding trials that British editors asked for. Ereky seems to have been unaware that what he claimed had already been done by Rouelle, Osborne and Chibnall, but he deserves attention because he had a clear idea of the advantages of fractionating forage.

The defects in Ereky's proposed machine were recognised by Goodall whose patent (1936) covered the use of three-roll mills, expellers and belt presses. He stressed that rubbing rather than cutting was essential as a first step in getting juice out of leaves, and that total disintegration into a puree was unnecessary and complicated subsequent handling. He was primarily interested in the medicinal merits of his product because it contained 'chlorophyl, vitamins, and ferments', but he also attached importance to the partly dewatered fibre because it could be economically dried in a current of unheated air to produce a winter feed for cattle. The use of juice from grass as a medicinal product was also patented by Schnabel (1938) who had for many years a somewhat mystical enthusiasm for pills made from protein-rich grass harvested before the formation of the first joint.

After this digression into pharmacology, interest reverted to the extraction of protein from leaves. Slade (1937) stressed the productivity of grassland and argued that, if it should be necessary to produce more of our own food in Britain, it would be better to fertilise grassland heavily, extract LP as a human food, and feed ruminants on the residue, than to plough up grassland. The proposition was included in a patent (Slade, Birkinshaw & ICI, 1939) which also covered many of the processes that had been used by Osborne and Chibnall, and that were regularly used in the separation of plant viruses from leaf extracts. There was more novelty in the suggestions that the protein curd should be compressed anaerobically and allowed to mature into a 'cheese', and that yeasts should be cultivated on the filtrate from the curd. Many patents have been allowed since 1939 although it is by no means clear, in the light of all the early academic work, and of the patents that have now lapsed, what points of any significance are being covered.

By 1939 the position had been fully defined: the quantity and extractability of protein in a few leaves was known, the advantages of extracting it were recognised, some methods for doing this were known, and so were the bulk properties of the protein. All that remained was to find out whether the production of leaf protein in bulk was practicable and whether the protein would prove to be as useful as theory suggested. I had started making plant virus preparations in 1934 and so gained experience in handling the normal proteins of the leaf. On the outbreak of war in 1939, many meetings were held in laboratories in Cambridge and elsewhere to discuss the ways in which scientific skill could be best used. Those of us who were interested in nutrition, remembering the effects of blockade in the 1914–18 war, suggested various projects for improving food production, with particular emphasis on protein. I suggested work designed to see whether new academic knowledge now made the bulk production of leaf protein feasible. Nothing happened immediately, but in the general ferment that followed the 'fall of France' in June 1940, I was asked to help the Food Investigation Board and ICI in a joint study of the problem. It seemed clear that there was little need for more work in the laboratory; instead, large-scale machinery had to be found or designed.

During 1940 and 1941, through the friendly cooperation of manufacturers and industrial users, many different hammer mills, screw expellers, sugar cane rolls, ball mills, rod mills, edge- and end-runner mills and dough-breakers (incorporators or Pfleiderers) were tried and all proved possible methods for making a satisfactory pulp, but they were all, for various reasons, unsatisfactory. Stamping mills of the type used to disintegrate ores were not tried because laboratory-scale work (Pirie, 1953) had not at that time demonstrated the possibilities of pulping by impact. It may be that this method deserves further study.

It was often difficult to persuade manufacturers of various pulping machines, tested for such a short time that an equilibrium temperature was not reached, that frictional heating was excessive. But it is obvious from the value of the mechanical equivalent of heat that, if 30 HP is consumed in making 1 t per hour of pulp containing 85 % of water, the temperature of the issuing pulp will be raised by nearly 30 °C. With such a power consumption, protein would therefore be coagulated *in situ* on a hot day unless one resorted to the further complication of cooling the pulper.

These tests on large-scale equipment, and some pilot-plant production of LP for palatability trials, started with ample encouragement from the relevant official bodies. There was a formal lunch in Cambridge in October 1940, attended by the Minister of Food, the Regional Commissioner, the Divisional Food Officer and others, at which LP soup, whale casserole, and other novelties were served. Questions were asked about LP in the House of Commons (e.g. Hansard 15/10/41, column 1370). By the end of 1941, euphoria had waned and the familiar official distrust of every unconventional proposal reasserted itself. There was a hilarious period in which I was told, at the same time and sometimes by the same people, that the idea of making LP as a human food could not possibly be practical, and that it was so important that I must be more secretive and not explain, to those whose pulpers I was testing, what was intended, lest the Germans should get hold of an important idea. The generous supply of food from the USA under 'Lend-Lease' finally made it obvious that there would be no need for British self-sufficiency in wartime.

Although government support for work on LP ended, considerable interest was still taken in the idea. In Britain, LP was discussed at several scientific meetings (e.g. Pirie, 1942*a*, *b*) and it was often mentioned in the popular press. Guha (1960) used LP from water hyacinth (*Eichhornia crassipes*) and other species during the Bengal famine in 1943, and Australians in a Japanese prisoner of war camp crushed 300 t of grass, rhododendron and passion fruit leaf between rollers so as to make an extract containing vitamins and protein (Smith & Woodruff, 1951). LP was used at the same time in the USSR as a source of vitamin A in human food, and as a protein supplement for calves, hens and pigs (Zubrilin, 1963, in Pirie, 1963). A return to the Winterstein (1901) method of drying and extracting with alkali was advocated in the USA (Sullivan, 1943, 1944). This process, as would be expected, has never been used on a large scale anywhere. In both Nebraska and California (Bickoff, Bevenue & Williams, 1947), LP was made by pulping fresh forage.

By 1948, interest in LP seemed to have disappeared in India, the USA and USSR. In Britain, the end of 'Lend-Lease' food supplies directed attention to the need for increased home production. There was, at the same time, more general awareness of the poor nutritional state of most of the world's inhabitants – especially those living in the wet tropics. Consequently, work on LP started again at Rothamsted and the Grassland Research Station (then near Stratford-on-Avon), supported by a grant from the Agricultural Research Council. Work at the Grassland Research Station was discontinued in 1953; work on LP at Rothamsted never formed part of the general program of the station but depended on special grants. Later improvements in technique and equipment depended mainly on a five-year grant from the Rockefeller Foundation in 1958, and a three-year grant from the Wolfson Foundation in 1965.

The early trials with large-scale equipment showed that hammer mills would pulp leaves satisfactorily if clogging could be prevented. In a conventional hammer mill, the charge remains inside until comminuted sufficiently to pass through holes in part of the casing. The dough that is made from pulped leaves clogs holes small enough to ensure adequate subdivision. This defect can be overcome by flushing the pulp through with water. That method has been advocated (Chayen, 1959; Chayen *et al.*, 1961), with the misleading designation 'impulse rendering', although it produces an inconveniently dilute extract. During discussion with Mr William Christy (of Messrs Christy & Norris, Chelmsford) in 1940, the idea evolved of making a hammer mill with a cylindrical pulping chamber two or three times longer than it was wide. This would have the intake at one end of the cylinder and discharge at the other, without any grid or screen. It would therefore be uncloggable: whatever went in would come out, whether properly pulped or not. To get the correct degree of pulping, an arrangement of beaters would be chosen that moved the charge through the pulper at a suitable rate. When work restarted in 1948, a 'coir sifter', normally used to separate coconut husk from fibre, was modified so as to work in this manner. During 20 years, pulpers have evolved to such an extent (Davys & Pirie, 1960; Pirie, 1971*a*) that their descent from a 'coir sifter' is no longer immediately obvious. One of the latest models is at the National Institute for Research in Dairying (Shinfield, Reading); earlier models are in various Institutes in Britain and also in India and Uganda. There is no need to describe this type of pulper because it is unlikely that others of similar design will be made.

Trials on a laboratory scale showed that 80 to 90 % of the amount of juice that could be expressed at any pressure, could be expressed in 5 to 10 s at 1.5 to 3 kg per cm^2 (150 to 300 kPa) if the layer of pulp is less than 2 cm thick. Obviously, power is wasted if there is relative motion between the charge and filtering surface while under pressure. Several presses meeting these specifications were made. In the first, a ram pressed down on pulp carried on a hinged loop of perforated conveyor; this moved in steps between strokes of the ram. At the instigation of the National Research Development Corporation this was patented (Pirie, Fairclough & Shardlow, 1958) although it had already been superseded by a press in which the pulp was carried on a rotating perforated table (Davys & Pirie, 1960). After considering more fully the principles on which successful expression of protein-rich juice depends (Pirie, 1959*b*), we (Davys & Pirie, 1965) made a reasonably satisfactory press in which pulp is pressed against a perforated pulley by a tensioned flexible belt. Twenty or 30 belt presses, with capacities ranging from 0.1 to 5 t of pulp per h have been made. The smallest unit, a scaled-down version of the one described in detail (Davys & Pirie, 1965), is the one most widely used. Its PVC-coated nylon belt is 5 m in circumference, 30 cm wide and passes round two pulleys 36 cm in diameter. One is perforated, and the other is forced outwards by springs able to exert 1 to 2 t. The belt moves at 2 to 4 m per min. There is no relative movement between the belt, the charge and the perforated pulley.

The tests made on large-scale equipment in 1940–41 and again in 1948–49 showed that existing screw expellers did not extract fibre-free juice satisfactorily from unpulped leaves. Nevertheless, the idea of extracting juice in one operation, in a piece of slow-moving equipment is obviously attractive – especially if use is envisaged in a less developed country where an animal might be the source of power. We (Davys & Pirie, 1963) therefore made a batch extractor in which 100- to 300 kg lots of leaf were pulped by a heavy, ribbed roller driven round a horizontal perforated bed. This extracted less protein than the other units, but used less power. Most of the LP used (p. 106) by Singh and his colleagues in the Central Food Technological Research Institute (Mysore) was made there with this machine; it is also used in Nigeria and Papua New Guinea. It has, however, several faults and that precise design would not be copied if simple units were being made now. But the basic principle – that LP will be most useful if it can be made with simple equipment by relatively unskilled people – is sound although it was then premature. In 1958, when the first unit was made, the concept of Intermediate, or Appropriate, Technology had not been formulated and organisations such as the International Bank for Reconstruction and Development and the grant-giving Foundations (e.g. Ford Foundation, 1959) were still obsessed by the idea that the world's food supply could be assured by intensive development of large-scale, sophisticated technology. This outlook did not change till about 1970. But for that technological obsession, more interest would probably have been taken in our work in the 1960s.

The factors that led to continued scientific interest in LP in Britain after 1948, whereas interest diminished in other countries, probably also explain the reawakening of commercial interest. Furthermore, the idea of using LP as food was at that time given some publicity in the popular press. The Angus Milling Co. Ltd (Aberdeen) started commercial production on a small scale. They separated grass into juice and fibre in a standard oil expeller: the dried fibre was 'Graminex Feed' and the coagulum from the juice was 'Graminex'. So far as I know, there is no publication on the process, and only one on the product (Carpenter, Duckworth & Ellinger, 1952), but a report written in 1951 by J. L. Dawson showed a comprehension of the principles involved, and the advantages that should result from commercial exploitation, that was unusual for that time. The defects of standard oil expellers were apparent to Powling (1953) and he designed a simplified expeller more suited to the job. This 'Protessor' is still being manufactured and is used in several institutes in Britain and elsewhere. Powling made LP from many different species of leaves, including leaves that are by-products; this was also the approach of Goodall (1950) who used sugar beet tops.

In several institutes (p. 132) unfractionated leaf juice is used as an animal feed. Most of the LP used in human or animal feeding trials was, and still is, made by heat coagulation; the reasons for choosing this method are discussed later (p. 47). The protein in leaf extracts can be precipitated, either completely or in fractions, in many other ways, e.g. by acidification or by adding salts, solvents miscible with water and water-immiscible polar solvents. The circumstances in which the use of these methods is advantageous will be discussed later (p. 65). Many leaf extracts coagulate spontaneously on standing for several hours at room temperature. There are few circumstances in which this would be a sensible method for making LP because of the autolytic and oxidative processes that happen at the same time.

3
Choice of crops and yields of extractable protein from them

With a forage, or other leafy crop, the part that is harvested is the part in which protein is synthesised, and not a part to which it is translocated. Because translocation losses are eliminated, a forage crop should, given a suitable climate and a similar level of husbandry, yield more protein per ha and year than any other type of crop. This expectation is borne out in practice. The advantage is even greater if the species used can regrow several times after harvesting so that the ground has a photosynthetically active cover throughout the growing season. That has been agreed for many years and was the basis of Slade's advocacy of LP production. In the late 1930s, when trustworthy amino acid analyses were beginning to appear, it seemed to many scientists (e.g. Chibnall, 1939) that the mixture of proteins in leaves should have good nutritive value. Even without the analyses this was probable because many nonruminant animals, i.e. animals with alimentary tracts in which there is little microbial synthesis of amino acids, live almost exclusively on leaves. Furthermore, the great metabolic and synthetic capacity of leaves forced us to assume that they contained a more extensive array of enzymes than any animal tissue. It seemed unlikely that the same amino acid excess, or deficit, would characterise all these different enzyme proteins. The distribution of amino acids in seeds and those other storage organs that have fewer metabolic activities is more likely to be nutritionally unfavourable. This expectation arises for purely statistical reasons, it does not depend on the invalid assumption, which has crept into some articles, that enzyme proteins are in some way 'better' than storage proteins – there are simply more different types of them in an active organ.

Though not always explicitly stated, the two cardinal propositions, that LP was potentially the most abundant source of protein, and that it should have good nutritive value, stimulated all the research outlined in the preceding chapters. Those who doubted the value of research on LP, or who were actively hostile to it, seem always to have paid too little attention to the basic propositions: they did not contest them, they simply disregarded them. Undoubtedly, difficulties could be foreseen. Otherwise LP production would not have been a research project. It may not be possible to maintain a useful green cover on the land in a practical farming system. It may not be possible to extract the protein without an economically unrealistic expenditure on machinery and energy. Other components of leaves may combine with, or damage, the protein in the course of extraction so that some essential amino acids are destroyed or made unavailable to nonruminants. And so on. These are matters on which research is still needed.

The technique of extraction

Once the feasibility of extraction had been demonstrated, attention could be turned to the selection of the most suitable leaves. The extractability of LP obviously depends in part on the vigour with which extraction is undertaken and the criteria that are used to distinguish extracted protein from protein still carried by the leaf fibre. When, as is usual in the laboratory, pulp is squeezed through a piece of cloth, the thickness and character of the weave are relevant, and also the amount of water added during pulping. Similarly,when pulp is centrifuged, the results depend on intensity and duration. Prolonged centrifuging sediments chloroplasts and their fragments, and both should be included in the category 'extracted LP'. With sufficiently prolonged grinding, e.g. in a motor-driven, end-runner mill, the leaf fibre can be so thoroughly disintegrated that it all passes into the supposed LP fraction. It is likely that nearly complete disintegration of the fibre explains occasional claims that nearly all the protein can be extracted, without enzyme digestion, in the laboratory. On the other hand, as already mentioned (p. 6), some methods of milling can attach protein to the fibre. For all these reasons, some standardisation of extraction technique is essential, and the techniques used should have at least a qualitative resemblance to techniques that would be feasible in actual large-scale extraction.

Early routine measurements of extractability were usually made by pulping in a domestic meat mincer, squeezing out the juice from a weighed amount of pulp by hand in a square of cotton cloth of the type used for pocket handkerchiefs, remincing the fibre with added water, and squeezing again. To minimise sampling errors, the water and nitrogen contents of the original leaf were measured on a sample of the first pulp. After measuring the volume of the mixed extracts, the amount of protein nitrogen and nonprotein nitrogen in it were measured by adding an equal volume of 10 % trichloroacetic acid (TCA) to a sample, centrifuging, and analysing the precipitate and supernatant. When working with leaves from species containing large amounts of nonprotein nitrogen, the precipitate should be resuspended in dilute TCA and centrifuged again. The extracted fibre was dried, weighed, ground, and analysed for nitrogen. From these measurements, the percentage of the total nitrogen appearing in each of the three fractions can be calculated, and the precision with which the operations were carried out can be assessed from the agreement between the mass of nitrogen in the sample of leaf taken and the sum of the masses of nitrogen in the three fractions. More nitrogen is precipitated from leaf extracts by TCA than by heating. The difference is usually 5 to 10 % but it can reach 20 % with extracts from very young leaves. The more rapidly the extract is heated, the more nearly the nitrogen precipitated by heating approaches the amount precipitated by TCA; the longer the interval between making the extract and adding TCA, the more nearly the amount precipitated by TCA approaches the amount precipitated by heating. These differences mainly arise because TCA precipitates the nucleic acid in ribosomes, and nucleic acid is destroyed by leaf ribonuclease during ageing or slow heating (Pirie, 1950; Singh, 1960).

Although much experience, and background information, was gained from measurements made by that method, domestic mincers have many defects. The individual plants in many crops are so large that at least 3 kg should be taken as a representative sample from a plot. Pushing that amount of material through a mincer is tedious and, with fibrous leaves, frequent stops may be necessary to disentangle fibre from the cutter. The juice liberated from lush leaves fills the barrel of the mincer and may, unless great care is taken, leak out backwards instead of coming out along with the pulp. It is difficult to make hand squeezing quantitative and consistent. Operators vary in persistence and in the extent to which they rearrange the charge within the cloth while squeezing. With financial support from the UK committee of the International Biological Program (IBP), a pulper and press were made that would give more systematic results (Davys & Pirie, 1969; Davys, Pirie & Street, 1969).

Careless operators, when grinding dry samples in a hammer mill, may leave some material in the mill at the end of a run and this will usually not have the same composition as the more easily disintegrated material that passes through the mill. This type of error must be guarded against in designing an analytical pulper. The possible error becomes smaller the larger the ratio of the mass of material pulped to the mass remaining in the pulper. At the end of a run, the IBP pulper contains 200–300 g of material. Experiments in which pieces of paper were put in from time to time along with the leaf showed that after 500 g had come through, equilibrium was established between emerging and retained pulp. That amount of pulp should therefore be discarded. The pulp produced closely resembles that coming from the large pulper and the percentage of the protein that is liberated into the juice is similar (Davys & Pirie, 1969).

The IBP pulper is a stepped drum 44 cm long, 27 cm in diameter at the feed end and 32 cm at the discharge end; the difference in diameter ensures an air flow in the direction of movement of the pulp. A rotor inside the drum carries 58 fixed beaters with a 2-mm clearance from the drum. In the laboratory, the rotor is driven by a 5-HP motor; the speed and direction of rotation are so arranged that the pulper can also be mounted on a 'Landrover' and driven from the power-take-off. It can therefore be used in the field. For safety the pulper is fed through a 5.5-cm diameter tube. Helped by a plunger so shaped that it cannot reach the rotor, feeding at up to 1.5 kg per min is possible. When used by operators who can be trusted to work safely, a wider entry tube can be fitted, thus permitting faster feeding. The outer drum is easily taken off so that all parts of the machine in contact with the crop can be cleaned.

The pulper is supplied with several pulleys so that it can be run at different speeds. For most purposes, 3500 rev per min is satisfactory; to ensure comparability in results it would be well if this speed were adopted as standard. There is, however, no advantage in running at such a speed that all the fibres are chopped short; on soft material it will therefore be interesting to have some measurements at smaller speeds. If pulping is inadequate, undamaged fragments of leaf will stand out as pale patches in the dark pulp.

The LP extracted on a pilot-plant scale for use in recent feeding trials (p. 108) is made from pulp produced by the IBP pulper, pressed in the small belt press already described (p. 13). That press is unsuitable for quantitative work. The pressed fibre from it is about 5 mm thick and it is exposed to 1.5 to 2.0 kg per cm². The IBP press simulates these conditions. Because the pulp is homogeneous, there is no need to work with a large sample. Because conditions at the edge of the material being pressed differ from those in the body of the cake, the larger the area of the cake the better. In the IBP press, 900 g of pulp is spread evenly on a cloth on a square, grooved platen 23 cm each way, the cloth is folded so as to make the area of pulp 450 cm² and another grooved platen is placed on top. The assembly is mounted with the grooves vertical and subjected to a pressure of 1 t applied by means of a bell-crank with a 40 to 1 ratio. These conditions closely resemble those in large presses and the dimensions were chosen so that the unit can be carried and, like the IBP pulper, can be put in a 'Landrover' and used in the field (Davys *et al.*, 1969).

The hard cake of fibre that is made by the IBP press cannot be quickly and easily broken up for reextraction. There is therefore no second extract to mix with the first, but mixing is as important as with hand squeezing because the composition of the juice coming out during the first few seconds differs from that coming out during the 2 to 3 min for which pressure is maintained. Pulp, fibre and extract are analysed as already described. Because pulping is more thorough in the IBP pulper than in a mincer, more protein is released from cells – especially with rather dry or fibrous leaves. On the other hand, because second extracts are not made, less of the released protein appears in the juice, and a correspondingly larger amount of the leaf nitrogen is retained in the fibre. From the soft, moist leaves usually studied, the IBP system extracts 20 to 30 % less protein than the older system. This discrepancy is not necessarily a defect; neither system reproduces faithfully the conditions of actual large-scale operation. Measurements made with these agronomic tools are intended to arrange crops and systems of husbandry in the order of their probable productivity; they cannot give precise information about the performance of large-scale equipment.

A crop harvested in a manner that does not bruise the leaves, does not deteriorate in a few hours in a cool climate. Deterioration is rapid after harvesting with a flail or 'precision chop'. It gets faster as the temperature increases. Batra, Deshmukh & Joshi (1976) measured the extractability of LP from seven crops kept at 28 to 35 °C for various times after harvesting. The yield from most of them was nearly halved after 9 h. They attributed most of this diminution to the evaporation of water from the crop so that there was insufficient juice to bring the LP out adequately. If the harvested crop had been sprinkled with water from time to time, there would presumably have been less loss of extractable LP. Until more is known about the cause and extent of deterioration, it would be prudent to harvest no more plots in an experiment than can be processed within an hour. For similar reasons, the condition of a crop early in the morning and late in the afternoon is not the same – especially in a hot, dry climate. Judgement is therefore necessary in agronomic work and the plots should be replicated so that similar ones can be taken at different times of day and be pulped after different amounts of delay. After pulping, speed is essential because protein immediately begins to coagulate onto the fibre and to autolyse in the extract. A delay of 2 h at 24 °C between pulping and pressing diminished the yield of protein from red clover (*Trifolium pratense*) to half (Davys *et al.*, 1969); Tracey (1948) and Singh (1962) got a similar diminution with wheat (*Triticum aestivum*). Pulp should therefore be pressed within minutes of being made.

Because of autolysis, there should be no delay in quantitative work during the operations of measuring the volume of juice coming from the 900 g of pulp in the IBP press, mixing it, taking a sample and precipitating it with an equal volume of 10 % TCA. Thereafter delay is immaterial. A crop, or potentially harvestable wild growth, can therefore be pulped, pressed and sampled with the IBP equipment at any site accessible to a 'Landrover' and the analyses can be completed at leisure in the laboratory.

Selection of species, and intraspecific variability

The tables that have been published listing the extractability of protein from the leaves of many species can be very misleading because extractability depends on many factors besides species. Early experience (e.g. Crook, 1946; Crook & Holden, 1948) justified the tentative generalisation that the greater the percentage of water in a leaf, and the greater the percentage of protein in its dry matter (DM), the greater the percentage of that protein that will be extractable from a given species by a given technique. This generalisation still seems to hold. A striking example was the observation (Arkcoll & Festenstein, 1971) that the yield of LP per m^2 was smaller from a plot of kale (*Brassica oleracea*) given phosphorus and potassium than from an unfertilised plot because, although the DM yield was increased 68 %, the amount of nitrogen in the DM was diminished by 56 %. Obviously, the generalisation holds only for comparisons between different treatments given to plots of the same species; species containing more protein do not necessarily extract better than others. It does however follow that there is a disproportionate advantage in harvesting leaves that are young, well manured and well watered. They will not only contain more protein, but more of that protein will be extractable.

Species differ in the extent to which maturity diminishes extractability; there are also differences between varieties. Crook & Holden (1948) could extract protein from one variety of strawberry (*Fragaria vesca*) but not from another. Byers & Sturrock (1965) compared five maize (*Zea mays*) varieties and found that the differences were not the same in successive years. Wheat varieties differ (Arkcoll & Festenstein, 1971). Differences in the date at which the haulm of different potato (*Solanum tuberosum*) varieties dies are well known and they affect the amount of protein that can be extracted from the haulm if that is taken as a by-product when the tubers are lifted (Carruthers & Pirie, 1975). Fat hen (*Chenopodium album*) shows extreme variability. Arkcoll (1971) singled it out as a species which continued to give good protein extraction after flower development; the variety used by Heath (1977) matured more abruptly than three other species with which it was compared; Carlsson (1975) found one variety yielding more than three times as much per m^2 as another, and 80 % of the protein was extractable from an Austrian variety, compared with 50 % with a Swedish variety. The yields of extractable protein from 15 varieties of fenugreek (*Trigonella foenum-graecum*) differed by a factor of five (Chandramani *et al.*, 1975*b*).

The ratio of protein nitrogen to nonprotein nitrogen differs between species; it also varies within a species with the conditions of husbandry, atmospheric temperature and possibly the time of day at which the leaf is harvested (Chibnall, 1939; Ostrowski *et al.* in Wallace, 1975). Because nonprotein nitrogen will be more difficult to use in commercial practice (cf. p. 118) than nitrogen in LP or associated with the fibre residue, species and conditions should be chosen in which nonprotein nitrogen is as small a fraction of the total nitrogen as possible.

Species that appear to extract well in laboratory conditions may be almost impossible to handle on a large scale, either because the extract is glutinous (e.g. comfrey *Symphytum asperrimum* and *Symphytum officinale*, or sweet potato *Ipomoea batatas*) or because it forms an intractable mass of froth. Some varieties of lucerne are difficult to handle for the latter reason. After working for a few hours the equipment may be difficult to find and almost impossible to manipulate. By contrast, some of the grasses that extracted badly when put through domestic mincers extracted well in the more violent conditions of the IBP pulper.

It follows from all this that a satisfactory basis for commercial LP extraction will not be established until there has been much more agronomic, or plant physiological, research. Sowing dates, fertiliser treatments, time of harvest and ability to regrow after harvest will have to be studied with many more species and varieties. There is also scope for plant breeding. The crop plant varieties used so far were all selected according to criteria, such as the ability to yield abundant seed, that may well be inimical to LP yield. The varieties most useful for LP production will probably have flowering delayed or prevented by sowing them at unusual times, or in unusual latitudes, or by genetic manipulation, or through use of growth regulators, so that senescence is delayed and there is a prolonged period of vegetative growth. No attempt has so far been made by plant breeders to use LP yield as a criterion. In selecting candidates for research on breeding, the points already made should be borne in mind, i.e. there should be abundant, lush, protein-rich leaf. Three more factors should be borne in mind. The leaves should not be carried on a very fibrous stalk or the energy needed for pulping will be excessive unless some form of scutching is possible; the leaf should be neutral or slightly alkaline, although acidity can be partly counteracted by pulping with added alkali, this complicates the process; the presence of tannins and phenolic substances diminishes protein extraction. There is enough in some leaves to prevent protein extraction completely (Bawden & Kleczkowski, 1945).

So as to give an impression of the potentialities of LP production, it is worthwhile surveying some of the results that have been published on protein extractability and yield although little information is often given about the antecedents of the plants used, and the most suitable varieties may not have been chosen. It is convenient to group possible sources of leaf arbitrarily into the categories:

Species used in conventional agriculture.

Leaves available as the by-product of a conventional crop.

Tree leaves.

Water weeds.

Miscellaneous unconventional species of leaf.

Conventional species

During the period in which extraction equipment was being designed and improved at Rothamsted, crops grown in the normal manner were used; latterly many experiments were on cereals sown two to four times more thickly than usual and fertiliser was used on some plots unusually liberally. This departure from normal farm practice was permissible because crops were harvested from late April until early June while still green and without a heavy head, there was therefore no risk of lodging. Our experimental plots were never irrigated although in this part of Britain there is a summer water deficit four years out of five. The results were collected in two papers (Byers & Sturrock, 1965; Arkcoll & Festenstein, 1971). So as to make use of winter sun and ensure an early start in spring, much of this work was on autumn-sown cereals. Proper timing of the first harvest is important because it affects the amount of regrowth. Species differ in the vigour of regrowth: with wheat it is excellent, with barley (*Hordeum vulgare*) less, with rye (*Secale cereale*) poor, and with maize negligible. When two, or sometimes three, harvests of short, leafy material were taken, the yield of LP was larger than the maximum yield obtainable from a single harvest; obviously, the amount of labour expended in harvesting was also greater. The land was then ploughed and sown with a variety of summer crops. The results of one such experiment are given in Table 1. More than 2 t of LP, i.e. 334 kg of protein nitrogen, was produced from the same piece of land within a year. Figure 1 shows the yields in successive years in experiments of this type. The increase in yield was the result both of increasing skill in harvesting crops at the correct time, and of improvements in the extraction equipment. The figure also shows the dependence of large yields on adequate summer rainfall.

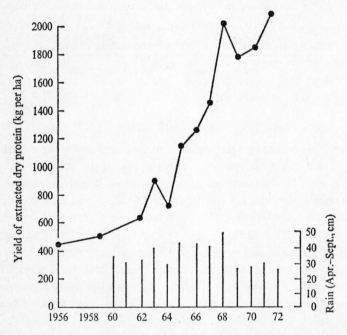

Figure 1. The yield of extracted protein (nitrogen × 6) from a succession of crops grown on the same plot without irrigation. Improvements to the extraction equipment, and increased skill in harvesting at the right time, contributed about equally to the increased yield between 1960 and 1968.

Table 1. *Yields of extracted protein (LP) in kg per ha (from Arkcoll &*
Festenstein, 1971)

Winter wheat sown 20/10/67,	harvested	1/5/68	819
	regrowth	18/6/68	284
Land then ploughed and sown with either:			

Mustard		Fodder radish	
First crop harvested		First crop harvested	
1/8/68	426	12/8/68	460
Second crop harvested		Second crop harvested	
29/9/68	490	15/10/68	461
Total yield	2019	Total yield	2024

Grass did not extract satisfactorily in early forms of equipment.
With improved equipment it became an excellent crop and had the
advantage of not necessitating repeated sowing. The well-known
beneficial effect of frequent cutting was shown clearly. Figure 2 shows
one such experiment on cocksfoot (*Dactylis glomerata*) harvested

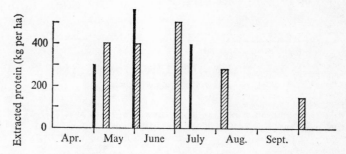

Figure 2. A comparison showing that the yield of extractable protein from single
harvests in 1969 of cocksfoot grass taken at different dates (the solid bar) was
much less than the sum of the yields from successive harvests (total yield was
1670 kg per ha) from the same plot (hatched bar).

from different plots at three different ages, compared with five harvests
from the same plot. The summed yield from five harvests was three
times the maximum from a single harvest, and 1969, as Figure 1
shows, was a dry year. Figure 3 illustrates the general phenomenon of
the decline in yield of extractable protein at about the time of
flowering.

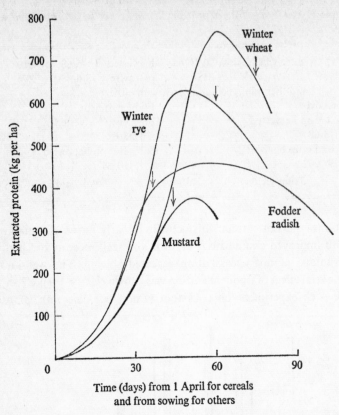

Figure 3. An indication of the manner in which the extractability of protein declines as leaves age, so that there is a definite optimum date for harvesting. Downward arrow indicates time of flowering of crop.

In all these experiments, increasing the amount of fertiliser nitrogen up to 264 kg per ha increased the yield of LP, although there was little increase in the yield of DM after about 132 kg per ha. Very large doses of nitrogen are, however, used inefficiently. Thus the largest increase in the yield of LP on going from 132 to 264 kg nitrogen was only 76 kg, i.e. only about 10 % of the additional nitrogen was recovered in LP. It is likely that it would not, as a rule, be economic to aim at these very large yields. The maximum yields from legumes depending for nitrogen on their root nodules, and without irrigation, were 1247 kg per ha with red clover harvested three times, and 1009 from lucerne harvested three times.

A group of agronomists in Reading (Britain) is extending research along similar lines (Heath, 1977; Heath & King, 1977). They find that true protein in cocksfoot, ryegrass (*Lolium perenne*) and fodder radish (*Raphanus sativus*) accumulates mainly in the laminae, and accumulation diminishes, or stops, when stem growth replaces laminar growth. The density at which a crop should be planted, the amount and timing of fertiliser application and the frequency of cutting, should all be managed so as to ensure maximum formation of leaf rather than stem. The inclusion of large amounts of stem in material from which LP is being extracted contributes little extra protein, and it acts as an absorbent mass within which the laminar protein gets lost. Heath contrasts the extra trouble and expense of repeatedly sowing annual crops, with the apparent economy of using perennial crops; he concludes that annuals give more flexibility, can probably be managed so as to give a more even supply of material throughout the season, and should in the conditions of commercial farming yield 840 kg of LP per ha. This is more protein than any other system of farming produces, and it does not take into account the value of the fibre residue and the nonprotein nitrogen in the juice.

With irrigation, but without nitrogen fertiliser, lucerne in New Zealand yielded 2 t of LP during seven summer months when processed in the IBP pulper. There was little difference in LP yield between four-week and five-week harvests; the latter gave larger yields of total DM (Allison & Vartha, 1973). This is an important point when the residue is thought of as a valuable fodder (p. 113). During the New Zealand winter, ryegrass (*Lolium multiflorum*) yielded 1.2 t of LP per ha from six harvests in 4.5 months. As at Rothamsted, the extra yield from more than 180 kg of fertiliser nitrogen per ha was gained wastefully (Vartha & Allison, 1973).

Other temperate zone conventional crops have been less thoroughly studied. Byers & Sturrock (1965) compared five clover (*Trifolium* and *Melilotus*) varieties. Apart from their ability to fix nitrogen, they seemed no better than the other species that were being used at the time; 40 to 50 % of their protein was extractable. Tares (*Vicia sativa*) came in the same range; they are a useful summer crop to sow after a winter cereal has been harvested. An unsuccessful attempt was made (Byers & Jenkins, 1961) to increase the yield by spraying the crop with gibberellic acid immediately after the first harvest had been taken. The merits of mustard (*Sinapis alba*) are apparent from Table 1. In another experiment (Arkcoll & Festenstein, 1971) a plot given 132 kg of nitrogen per ha produced 10.6 kg of extractable protein per day during August and September. Rape (*Brassica napus*) produced less. Sunflower (*Helianthus annuus*) produces very large yields of both DM and protein if given enough nitrogen, but this protein does not extract well from mature plants (Lexander *et al.*, 1970). These measurements were made on leaves that had been frozen. It is possible that freezing coagulated some of the protein and so made it unextractable, but earlier (unpublished) experiments at Rothamsted, made on material pulped while fresh, were essentially similar unless very young plants were used.

There is research in at least nine institutes in India, and so much of the reliable quantitative agronomic information comes from that country, that a digression on the reasons for this interest may be relevant. In 1959, in casual conversation, I heard that Jayaprakash Narayan was in London. Because I had criticised some aspects of the Bhoodan movement in India, I was urged to discuss with him the place of technology in a rationally run society (p. 13). He showed so much interest in research on LP that I sent him some papers, which he gave to Jawaharlal Nehru, who asked the Council for Scientific and Industrial Research to look into the matter. I was invited to India at the beginning of 1961; research started soon after that in the National Botanic Gardens in Lucknow and in the Agricultural College (now University) in Coimbatore. Soon after the start of the IBP, research in Coimbatore was complemented by research in Aurangabad and Calcutta, using IBP equipment donated by the UK committee of the IBP (Pirie, 1976*b*).

Lucerne, given irrigation water and some nitrogen fertiliser, yielded more than 3 t of LP per ha in a year at Aurangabad in spite of attack by aphids (Dev, Batra & Joshi, 1974). More than half was from six to eight harvests taken in the six months October–March. Several additional points were made in that series of experiments: the yield was increased by sowing in rows that were 23 rather than 46 cm apart. The yield of LP was increased 47 % in the first year, and 60 % in the second, when lucerne was irrigated during summer (Savangikar & Joshi, 1976). In some experiments, growth regulators such as 'Simazine' increased the protein content of leaves: 'Simazine' did not increase the yield of LP from lucerne (Dev *et al.*, 1974) or hybrid napier grass (*Pennisetum typhoideum* × *P. purpurea*) (Gore & Joshi, 1976*b*). Earlier work by the same group (Joshi, 1971) had shown that inoculating lucerne seed with rhizobia increased the yield of LP by 12 to 19 %. In some parts of India, lucerne does not survive well; it is treated as an annual rather than a perennial crop. Until the reasons for this poor survival have been found, and possibly corrected, lucerne lacks one of the advantages that it has elsewhere.

Byers (1961) noted good extraction of LP from cowpea (*Vigna sinensis* or *Vigna unguiculata*). It has many merits: LP made from it has a pleasanter flavour than LP made from lucerne, and its associated rhizobia do not waste 40 to 60 % of their metabolic energy, as those associated with soy and lucerne do, in making hydrogen rather than fixing nitrogen (Schubert & Evans, 1976). Cowpea responded consistently to inoculation (Deshmukh & Joshi, 1973). Nevertheless, it was usually given farm-yard manure and 20 kg nitrogen per ha; more nitrogen gave no extra yield. When grown in summer with that treatment, it yielded 895 kg of LP per ha from three harvests during 80 days, i.e. it accumulated 11.2 kg of LP per day (Deshmukh, Gore, Mungikar & Joshi, 1974). An earlier experiment (Deshmukh, unpublished) gave 16.9 kg per day; if this can be repeated it will be a world record for LP production. These yields should be borne in mind if any attention is being paid to a discouraging forecast of the potentialities of LP (Anonymous, 1965) based on the ludicrous daily yield from cowpea of 0.5 kg per ha.

Tetrakali, the tetraploid form of *Phaseolus aureus*, grows well in the wet season in West Bengal where it is used as fodder. It produces LP at rates that would correspond to an annual yield of 1760 kg per ha if growth had been maintained by irrigation during the dry season. The crop was studied for three years with different harvesting rhythms and spacings of the plants. Bagchi & Matai (1976) conclude that it is an admirable crop which, on better soil than they have access to, would give large yields. Berseem (*Trifolium alexandrinum*) does not yield as well as cowpea, lucerne or tetrakali, perhaps because extractability declines unusually rapidly as the crop matures (Shah, Zia-ur-Rehman & Mahmud, 1976; Mungikar, Tekale & Joshi, 1976a). This legume may, however, become important because it is already familiar in India and Pakistan, and its soft leaf favours economical extraction. Yields from hybrid napier grass varied from 2.2 to 0.6 t per ha and year in spite of liberal use of fertiliser nitrogen (Gore, Mungikar & Joshi, 1974). The variation probably depends on the precise interval between successive harvests; this crop quickly becomes fibrous which, as already explained (p. 23), wastes energy during extraction and diminishes extractability. The yield of DM is greatest with 50-day intervals between harvests, the yield of LP is greatest with 18- to 23-day intervals. The largest recorded annual yield of LP was 0.98 t per ha, but reasons are given for thinking that 2 t would be attainable. Yields in West Bengal (Matai, Bagchi & Chanda, 1976) were consistently smaller than those in Aurangabad. Yields from the three grasses *Bracharia mutica*, *Cenchrus glaucus* and *Panicum maximum*, grown further south in Coimbatore (Chandramani *et al.*, 1975a; Balasundaram *et al.*, 1975), depended on the frequency of cutting. *P. maximum* gave 1.3 t of LP per ha in a year when cut every 30 days, with a daily yield of 15.8 kg during one month, whereas the yield was only 0.5 t when the interval between harvests extended to 45 days.

Although the efficacy of green manuring is doubtful in regions with high soil temperatures, the technique is widely used. Species are chosen which give an abundant yield of nitrogen-rich foliage quickly; these are the basic characteristics of a good source of LP. Where green manuring is considered desirable, it would seem better to plough in only the two by-products from LP extraction. Some of the crops already discussed, e.g. lucerne and mustard, are used, so are a few, e.g. kudzu (*Pueraria phaseoloides*) (Byers, 1961) and *Sesbania sesban* (Gore & Joshi, 1976a), that have been tried as sources of LP. The latter extracted well but was not thought preferable to lucerne or cowpea. However, the extensive range of tropical legumes commonly used as cover crops, green manures, or hedge plants has not yet been adequately studied. This is a promising group of species because the techniques of husbandry are already familiar.

By-product leaves

The idea of using as a source of LP, leaves that are the by-product of some conventional crop has an obvious appeal. Unfortunately, the range of potential sources is limited. When a dry seed such as maize, rice (*Oryza sativa*) or wheat is harvested, the leaf is withered. Sugar cane (*Saccharum officinarum*) is a possible source. The tops are now usually burnt because, by removing trash, this makes hand-cutting easier, and by removing snakes and scorpions, pleasanter. With mechanised cutting these advantages of burning lose some cogency; furthermore, burnt cane deteriorates rapidly and must therefore be harvested and processed quickly or there will be loss of sugar. However, even in regions that are so humid that the tops are still green and fairly moist at the time of harvest, they seldom contain more than 1.2 % nitrogen (in the DM). That, however, corresponds to at least 0.5 t of protein per ha; Balasundaram *et al.* (1974*a*) extracted 108 kg of LP per ha. Cane tops are so fibrous that the energy expenditure on extraction would be large. If they were used as fibrous material with which to mix leaves such as those from sweet potato and water hyacinth, which are so soft that they are not easily handled in existing machines, some protein would be extracted from the cane tops as well.

The area already devoted to cassava (tapioca, manioc, or yuca, *Manihot esculenta* or *Manihot utilissima*) is so large, and there are such extensive plans for increasing production both for food and for industrial alcohol, that thorough investigation of its leaf would be worthwhile. Byers (1961), Singh (1964) and Balasundaram *et al.* (1974*a*) extracted little protein from it. But another paper by the last authors (1974*b*) put cassava in their 'category 1' in a classification of 212 species. They do not state clearly that the leaves used were taken at the time of normal harvest of the tubers. Efficient extraction equipment was not used in these experiments: that is an important point because cassava leaves when mature are rather dry and tough. Other by-product leaves, from which it is not likely that protein could be economically extracted, are those cultivated for the sake of essential oils or perfumes; they are usually subjected to drying or steam distillation, and these processes coagulate the protein *in situ*.

In the temperate zone the two most abundant leafy by-products are sugar beet (*Beta vulgaris*) and potato. The former is largely wasted in Britain, though not in Germany or Poland; the latter is wasted in almost all countries. An argument sometimes advanced for ploughing-in these by-product leaves is that they have manurial value. That is partly true. But Widdowson (1974) showed that the extra yield of barley in the following year, attributable to ploughing-in 130 kg of nitrogen in sugar beet tops, would have been given by 8 kg of fertiliser nitrogen. Goodall (1950) made a product containing 38 to 40 % protein on a commercial scale. At Rothamsted, LP containing 60 % protein was made (Pirie, 1958). The maximum yield quoted was 500 kg per ha, but the yield obviously depends on weather and on the date of harvest. However, even if the yield in practice were only half our experimental yield, the 193 000 ha on which sugar beet is grown in Britain would produce 48 000 t of extracted protein as well as a fibre residue that would be more attractive as cattle fodder than the untreated tops. More research is needed to find out whether the usual technique of separation (Morrison & Pirie, 1961) removes oxalate adequately from the LP.

As a safeguard against blight, and to facilitate tuber-lifting later, potato haulm is usually destroyed mechanically or with a herbicide at the beginning of September. It would not be difficult to design a haulm-harvester to work on potatoes ridged up in the usual way; it would be more convenient to harvest haulm from a level field, and there is some reason to think that the main merit of ridging is simply the suppression of weeds. If so, weeds could be controlled with herbicides, and the haulm could be harvested like an ordinary forage crop one or two weeks before the tubers were to be lifted. The main reason why potato haulm is disregarded in Britain as a forage crop is fear that stock will be poisoned by solanin and other glycoalkaloids. These are separated in the 'whey' when LP is made. It has been made regularly at Rothamsted since 1952, and has also been made in India, Pakistan and Poland.

The yield of LP depends on the potato variety and the date on which the haulm is taken (Carruthers & Pirie, 1975). Some early varieties yield 600 kg per ha, maincrop can yield 300 in late August, but yield diminishes to 100 or even less by mid-September. Once the idea has been accepted that something valuable could be extracted from potato haulm, prudent farmers would probably harvest haulm a little earlier than usual as a precaution against blight, but tuber weight is still increasing at the beginning of September, it would therefore seldom be worthwhile taking the haulm early simply because of the increased yields of LP; a trend towards earlier lifting is, however, foreseen in Britain (Ivins, 1973) and is already in practice in Australia (Sale, 1973). In Britain, early potatoes occupy 30 000 ha and maincrop 230 000. It is reasonable to conclude that about 50 000 t of LP could be extracted from the haulm. Soon after potatoes were introduced into Britain, Sir Thomas Overbury is said to have remarked: 'The man who has not anything to boast of but his illustrious ancestors is like a potato, the only good belonging to him is underground'. The sentiment is admirable, but we may be taking the simile too literally.

At one time, peas (*Pisum sativum*) for canning and freezing were harvested on the vine and the whole mass was taken to a factory, the peas were separated and the vines carted back to the farm. Mobile viners now go into the field and drop the vines as they go along. The old arrangement was well adapted to protein extraction. From pea haulm grown on an experimental plot we got 600 kg of LP per ha; this is more protein than is present in the peas. Pea haulm collected from a factory 30 km from Rothamsted did not extract so well as fresh haulm – perhaps because there was an interval of about three hours before the bruised and battered haulm was pulped. The percentage of haulm nitrogen that was recovered as protein nitrogen varied from 22 to 47 % (Byers & Sturrock, 1965). If efforts were made to give up vining in the field and revert to the old method of vining in a factory, protein could be extracted from the haulm without delay. However, peas may not ultimately be a useful source of LP; there is little leaf on some new varieties.

When maize is allowed to ripen completely, the leaves are too dry and depleted of protein for satisfactory extraction. From some varieties, harvested at the end of August, only 20 % of the protein was extractable (Byers & Sturrock, 1965). When harvested earlier, i.e. at the sweet-corn stage, nearly half the protein was extractable and the yield of extracted protein reached 480 kg per ha. The area devoted to sweet-corn in Britain is too small for it to be an important source of LP.

Vegetable discards in the USA in 1948 had a wet weight of 4 Mt (Willaman & Eskew, 1948); the weight in 1958 was 21 Mt (Oelshlegel, Schroeder & Stahmann, 1969), containing an estimated 0.4 Mt of protein. There have been no similar estimates recently in Britain: in 1951 the wet weight was 0.5 Mt, the weight sometimes quoted now is 4.6 Mt (Raymond, 1977). The area devoted to open-air vegetables in England and Wales is 170 000 ha, i.e. intermediate between sugar beet and potatoes. The brassicas alone (Brussels sprouts, cabbage, calabrese, cauliflower, sprouting broccoli, etc.) occupy 60 000 ha. The weight of discarded material is not recorded but is probably more than 200 000 t containing more than 3000 t of protein. Material discarded in the field would be fresh and worth extracting; discards at the retail level may possibly not be worth collecting. Much of the leafy waste in the field is already being collected and some market gardeners pay to have it disposed of.

Tekale & Joshi (1976) point out that vegetable growing gives Indian farmers a better income than other types of farming, and that recent improvements in vegetable varieties are as sensational as those widely publicised with the new cereal varieties. A survey by FAO (1971) found that vegetable consumption in India was almost the smallest in the world. It is therefore likely that market gardening will soon increase greatly, and it is fortunate that several by-product leaves have already been studied there. Yields of LP from the brassicas were 90 to 160 kg per ha (Matai *et al.*, 1973; Deshmukh *et al.*, 1974; Tekale & Joshi, 1976). Other useful by-product leaves from market gardens were chicory (*Cichorium intybus*) (Mahadeviah & Singh, 1968), sweet potato (Byers, 1961; Balasundaram *et al.*, 1974*a*; Deshmukh *et al.*, 1974), beetroot (Tekale & Joshi, 1976), and radish (Tekale & Joshi, 1976). Hussain, Ullah & Ahmad (1968) estimate that 60 000 t of LP could be made from by-product leaves available in Pakistan, and that half of it could come from radish. The potentiality of groundnut leaves (*Arachis hypogaea*) depends on the rainfall just before the seed is harvested. In Ghana (Byers, 1961) it extracted reasonably well, but in Nigeria, (quoted in Pirie, 1971*a*) and India (Jadhav & Joshi, 1977) it extracted poorly when the leaves were taken at the normal time of harvest.

Unpublished experiments at Coimbatore showed that leaves gathered from cotton (*Gossypium hirsutum*) after the second picking of the bolls yielded 510 kg of extracted protein per ha. After the third picking the leaves were too dry to extract well. The methods now used for mechanised cotton picking depend on chemically defoliating the plants; though it may not be easy to combine the collection of leaf with collection of bolls, the area devoted to cotton is so large that the attempt would be worth making. Ramie (*Boehmeria nivea*) is now being grown more extensively than hitherto because of the increased cost of artificial fibres and the exceptional resistance of ramie fibre to weathering and water damage. The leaf is scutched off before retting and contains 2.25 to 4.5 % nitrogen according to age and level of manuring (Squibb, Mendez, Guzman & Scrimshaw, 1954). Judging from appearance and feel it should extract well, and in (unpublished) experiments at Coimbatore 63 % of the protein was extractable; less protein was extracted in other experiments (Ghosh, 1967) unless alkali was added. Some other fibre plants have similar nitrogen contents but a harsher feel; this is in agreement with the observation (Byers, 1961; Matai, Bagchi & Roy Chowdhury, 1971) that only 30 to 40 % of the protein in sunn hemp (*Crotalaria juncea*) and only 20 % of the protein in roselle (*Hibiscus sabdariffa*) is extractable. The latter has an acid leaf and would therefore probably extract better if neutralised. Kenaf (*Hibiscus cannabinus*) was originally grown as a tropical fibre plant; there is now increasing interest in it as a source of paper pulp in New Zealand (Withers, 1973), Russia and the USA. It has acid leaves and is presumably similar in other ways to roselle. Some of the jutes (e.g. *Corchorus* sp.) have mucilaginous leaves that give an extract that is difficult to handle; this should be remembered lest it be assumed that every protein-rich leaf could be a source of extracted LP.

Protein extraction from the other fibre plants is less likely to be successful. Sisal (*Agave sisalana*) is acid. Judging from experience with protein from other leaves, neutralisation after the pulp has been made will not make as much protein soluble as would remain soluble if the leaf were scutched with dilute sodium carbonate so that neutralisation and disintegration are simultaneous. There is no reason to think that this neutralisation would damage the fibre. It may also be possible to recover protein from abaca (*Musa textilis*) leaves, but experience with banana (*Musa sapientum*) (Byers, 1961) is not encouraging. New varieties of abaca are being produced as sources of paper pulp; it may be that some of them do not contain so much of the phenolic material that is probably (p. 83) the reason for poor protein extraction. Some of the many varieties that are being tested as sources of paper, in various parts of the world, could be useful sources of by-product leaf.

Tree leaves

When first confronted with the idea of making LP, most people think immediately of tree leaves. Nevertheless, there has been very little work on them. A few points are obvious. The autumn leaf fall is too dry and contains too little protein to be a likely source; extraction is equally unlikely from the hard, dry leaves of many tropical trees; it would usually be difficult to mechanise collection of the leaf from large trees. There are several ways of coping with the last difficulty. As knowledge increases about the normal process of leaf-shedding it should become possible to induce it in the waste branches that accumulate during forestry, and thus separate leaves from twigs. Complete separation may not be necessary. In Russia, *muka* is made as fodder by beating the branches in a mill and separating the leafier fraction (Young, 1976). Trees could be grown as large hedges and trimmed mechanically from time to time – as is already done with tea, again in Russia. Coppicing is the most probable technique. It is already used for making paper pulp and silage in the USA (Herrick & Brown, 1967; Dutrow, 1971). Once the cycle is established, the many stems growing from sycamore stumps are harvested mechanically when they are 2 to 5 cm thick and stripped of their leaves. Unfortunately, sycamore does not extract well. Elsewhere, bamboo is treated in a similar way and protein has been extracted from its leaves.

The productivity, protein content, and distribution of amino acids were measured on 10 species of trees growing in Sweden (Siren, Blombäck & Alden, 1970; Siren, 1973). Natural growth contains 1 to 2 t of protein per ha; with intensive cultivation and fertilisation the quantity could be increased to 6 to 7 t; the range of protein contents of the leaves was similar to that of conventional crop plants. This was also the conclusion come to in Romania (Nedorizescu, 1972). Tree leaves are therefore a valuable potential source of animal fodder; there is less evidence that they could be a source of extracted protein. Crook & Holden (1948) commented on the toughness of some and the consequent difficulty of getting out any juice. Casual experience at Rothamsted since then confirms that judgement; elder (*Sambucus niger*) is the only local species from which we have extracted LP satisfactorily. There have also been casual observations on tropical trees. Byers (1961) got a small yield of good material from *Leucaena glauca*; in the Philippines a related tree legume (Bayani or ipil-ipil) is being tried. Leucaena deserves study because it grows so vigorously as to be a pest in some parts of the world, and it is of uncertain value as an unprocessed cattle fodder because it sometimes contains enough mimosine, a toxic, soluble amino acid, to cause loss of hair and hooves, and thyroid damage (Jones, Blunt & Holmes, 1976). Judging purely from appearance and texture, *Gliricidia maculata* seems to have potentialities, but in one trial, extraction was poor (Balasundaram *et al.*, 1974*b*). Choosing suitable trees for study is a matter for people familiar with the local flora, and with the physical characteristics of a leaf that will extract well. It may be worthwhile emphasising that nothing would be gained by extracting protein from a tree leaf if it is already being eaten as a familiar green vegetable (pp. 42, 84).

Although little is known about LP from tree leaves, some space has been devoted to them because a food-producing tree crop would be the ideal replacement for tropical rain forest. It is now obvious that ecological disaster follows attempts to cultivate annual plants in regions where there is frequent intense rain. Unless the soil is protected by the continuous presence of the roots of perennial plants, erosion is a serious risk. The trees usually thought of as replacements for natural rain forest produce an exportable commodity such as rubber or palm oil; it would be advantageous if some of the trees produced protein for local consumption instead.

Water weeds

Although many popular articles suggest that LP will be made from the mixed growth of weeds on land, this is unlikely. Several of the individual species in the mixture may ultimately be cultivated (p. 44), but the mixed growth on untended wasteland will contain little protein because it will not have been manured, and the land will probably be rough or in some other way unsuited to mechanical harvesting – otherwise it would not be wasteland. The situation is different with water weeds. Obviously they do not suffer from drought, and they are often abundantly manured. When this abundance is supplied by sewage, or the run-off from too heavily manured farmland, it is elegantly called eutrophication. The excessive growth of water weeds is harmful for many reasons; it increases the waste of water by evaporation in hot countries, interferes with flow in irrigation ditches, interferes with navigation and is sometimes a health hazard. The explosion of weeds in Lake Volta (Ghana) has encouraged snails and thus increased the incidence of schistosomiasis, or bilharzia (Derban, 1975). In spite of the value of a moderate growth of weed as a source of fish food (e.g. Green, Corbet, Watts & Lan, 1976), a great deal of effort is expended on attempts to batter weeds to death *in situ*, kill them with herbicides, or establish some system of 'biological control'. These methods may control growth but they do not affect eutrophication. The killed plants rot in the water and most of the elements they contain, and the excreta of an unharvested agent of 'biological control', are likely to return to the water. When herbicides are used, they and their breakdown products may remain in the water also and so make it unsuitable, or less suitable, for irrigation. If water weeds were removed from the water they infest, that water would not only be freed from infestation but would be depleted of the elements causing eutrophication. This would economically convert a problem into an asset.

The older literature on the use of water weeds was reprinted by Little (1968); many more publications have appeared since. Nevertheless, a recent Ministry Bulletin (Robson, 1973) did not mention use as a means of control. At a seminar in India (Varshney & Rzoska, 1976), only 5 papers out of 51 dealt with use. It has been stated that, worldwide, attempts at control are costing 500 M£ annually; the annual cost in Britain is 3 M£. There are signs that use, as an alternative to destruction, will soon get more attention. It formed part of the title of a UNESCO symposium (Mitchell, 1974), has been stressed in several papers (e.g. Bates & Hentges, 1976) and forms part of the program of a Commission of the US National Academy of Sciences (1976).

Processes that drag weeds out of the water, either to be used or allowed to rot, remove nutrients from the water, where they cause eutrophication, on to the land where they increase fertility. The amount of nitrogen, phosphorus and potassium removed depends on the luxuriance of growth, i.e. on the amount of nitrogen, phosphorus and potassium present, and on the ratio of the concentrations of these elements in the water. Thus Wooten & Dodd (1976) say that water hyacinth removes almost all the nitrogen, but little of the phosphorus, whereas Ornes & Sutton (1975) say that it removes 86 % of the phosphorus. In one water, nitrogen was presumably limiting growth, and in the other water, phosphorus. Boyd (1976) stresses the merits of water weeds as a means of removing from water the elements on which excessive weed growth depends.

Water hyacinth is the most abundant, troublesome and decorative of the weeds. It is said to cover 200 000 ha in India, and similar areas elsewhere. The total area is probably more than 1 000 000 ha with an annual growth of 10 to 30 t DM per ha. The DM of the whole floating plant contains about 2.5 % of nitrogen, the separated leaves contain up to 5 %. An impressive amount of protein is therefore potentially available. Unfortunately, it does not extract readily unless alkali is added (Ghosh, 1967; Taylor, Bates & Robbins, 1971; Matai, 1976); extraction is being commercialised in the Philippines (Monsod, 1976). Pieterse (1974) reviewed the literature on the biology and use of water hyacinth in a perhaps unreasonably pessimistic manner. Two other floating weeds, Nile cabbage (*Pistia stratiotes*) and the fern *Salvinia auriculata*, extract as badly as water hyacinth at their natural pH (about 6) (Byers, 1961; Matai *et al.*, 1971; Matai, 1976; Rothamsted Experimental Station, unpublished); the effect of adding alkali to them has not apparently been tried. It would be easy and economical to collect these floating weeds with equipment mounted on a barge, process the extract on it, and discharge the soluble material. The compact extracted fibre and LP would be off-loaded less often than harvested weed would have to be. That manner of working would obviously not control eutrophication as effectively as total weed removal.

Protein appears to extract more easily from rooted than from floating water weeds: there is no obvious physiological basis for this distinction. From mixed weeds collected in Hertfordshire, 47 % of the protein was extractable (Pirie, 1959c); from a water lily (*Nymphaea lotus*) in Ghana 40 % (Byers, 1961), and from one in Alabama (*Nymphaea odorata*) 61 % (Boyd, 1968). Boyd (1968, 1971) lists several other species that extract well; *Justicia americana* yielded 300 kg of LP per ha when harvested in May or June. Some of these water weeds regrow with disconcerting readiness after being cut as a means of control. Unfortunately the reeds (*Typha* and *Phragmites*), which produce very heavy crops in suitable regions (Dykyjova, 1971), do not extract well. Matai (1976) extracted from *Chrozophora* sp. and *Allmania nodiflora* 37 % of the leaf nitrogen in the form of LP.

No measurements seem to have been made of the extractability of the protein in the algae or angiosperms growing in sea water. If the problem of collection could be solved, the angiosperms seem to have potentialities. Thus turtle grass (*Thalassia testudinum*) is as productive as most conventional crop plants; its rhizosphere fixes 100 to 500 kg of nitrogen per ha and year (Patriquin & Knowles, 1972); growth is therefore probably limited by phosphorus deficiency. Most of the algae (seaweeds) contain too little protein to make extraction probable. Those species that are rich in protein, e.g. *Ulva* and *Porphyra*, are already used as food.

Miscellaneous unconventional species of leaf

It was at one time reasonable to examine many of the plants found growing wild so as to characterise the types of leaf from which LP could readily be extracted. This enabled a set of expectations to be formulated which could guide the search for species that might be better than conventional crop plants as sources of LP. This was the policy lying behind the work of Crook & Holden (1948) on 26 species, Byers (1961) on 76, Singh (1964) on 12, Devi, Rao & Vijayaraghavan (1965) on 16, Nazir & Shah (1966) on 25, Gonzalez, Dimaunahan & Banyon (1968) on 42, Sentheshanmuganathan & Durand (1969) on 12, Lexander *et al.* (1970) on 24, Deshmukh & Joshi (1969), Dev & Joshi (1969) and Gore & Joshi (1972) on 61, Garcha, Kawatra, Wagle & Bhatia (1970) on 20, Matai *et al.* (1971) on 20, Subba Rau, Ramana

& Singh (1972) on 19, Matai & Bagchi (1974) on 34, Carlsson (1975) on 71, and the species listed between 1948 and 1969 in Rothamsted Annual Reports. Balasundaram *et al.* (1974*b*) examined 212 species, but gave details about the most promising ones only. There is considerable overlap in the lists of plants in these publications: about 300 species have been studied. Various methods were used: the yields given in one paper cannot therefore be compared with those in another, and none gives precise information about the yields that would be given by a large-scale extractor. Useful as this work was, it belongs to what may be called the Natural History of the subject. Publication of more measurements of this unsystematic type does not seem to serve any useful purpose, and it is a pity that editors still permit it. Except in special circumstances, the only measurements that are useful today are those made in a standardised manner, e.g. with the IBP unit (Davys & Pirie, 1969; Davys *et al.*, 1969), or with a comparable unit which has been tested on a crop from which the extractability of LP is well known. When measurements are made on plants picked from odd sites in the neighbourhood, they demonstrate only that a certain species may be worth investigating more thoroughly. Selected species should then be deliberately cultivated on land of known manurial status, harvested at a known age, and the yield from a known area should be measured. Information less comprehensive than that is now scarcely worth publishing.

Of the quarter of a million species of flowering plants, the conventional crop plants may well be the most suitable when they are used in conventional ways and are grown in climates similar to those of the region in which they originated. Many analyses of leaves from wild plants (e.g. Slansky & Feeny, 1977) show that they may contain more nitrogen than the leaves of cultivated plants grown in the same conditions. They are often more resistant to attack by pests and predators. This resistance is often correlated with the presence of toxic substances, but these would probably be separated from LP during its preparation.

Hitherto, we have decided which crops should be grown, and have then extended their range into increasingly improbable regions. The alternative approach is to examine the species that already grow well in a region to see whether some use, in this context the extraction of LP, could be made of them. Attention should first be given to species that are already cultivated on a small scale, and that are not used as sources of leaf. For example: quinoa (*Chenopodium quinoa*) is cultivated for its seeds in the Andes, its soft texture makes protein extraction economical (Pirie, 1966*a*), and it was one of the species investigated in detail in the laboratory by Carlsson (1975). Mustard (p. 25) is cultivated mainly as a green manure in Britain; its relative *Brassica nigra* is an oilseed crop in India. If harvested young it re-grows well and gives excellent yields of LP (Matai *et al.*, 1973, 1976). Turnip (*Brassica rapa*) was disappointing; it does not regrow well (Matai *et al.*, 1973) and the yield of LP is poor when it is allowed to grow until there is a saleable root so that the leaves can be considered a by-product (Matai *et al.*, 1976). *Brassica carinata*, which is used as an oilseed crop in Ethiopia, appears to have potentialities in Texas as a source of LP (Brown, Stein & Saldana, 1975). A more unexpected suggestion is that *Tithonia tagetiflora*, usually cultivated because of the beauty of its flowers, should be used; it regrows well after harvesting and responds vigorously to manuring (Deshmukh *et al.*, 1974; Mungikar, Batra, Tekale & Joshi, 1976*b*).

There seems to be little substance in the argument sometimes advanced that LP would be more acceptable if it were known to be made from leaves that are already familiar as green vegetables. So few communities eat vegetables to the extent that is both desirable and physiologically reasonable (Pirie, 1975*a*, 1976*a*, *c*) that nothing would be gained by extracting protein from leaves that could be eaten in the natural state by adults. The situation is different with infants. The partial removal of oxalate is advantageous and, when the rest of the diet is bulky, there may not be room in their small stomachs for the quantity of a leafy vegetable that would be needed to meet their β carotene (pro-vitamin A) requirement; the equivalent amount of LP is easily accommodated (p. 92). The familiar green vegetables are no better as sources of LP and β carotene than some of the species discussed in earlier sections of this chapter. Instead of investigating the extraction of LP from familiar vegetables for infant feeding, it would be better to see whether some of these other species, when cultivated with skilled husbandry, could be used as leafy vegetables, as young lucerne and oats (*Avena sativa*) already are in some places.

Nevertheless, Gunetileke (quoted in Pirie, 1971*a*) claims impressive yields from *Basella alba*, and Fafunso & Bassir (1976) extracted *Corchorus olitorus*, *Solanum incanum*, *Solanum nodiflorum* and *Talinum triangulare*. The yields of LP were, however, no greater than those from forage crops.

A special section of this chapter was devoted to water weeds because they already grow to such an extent that they interfere with the use of the water and so have to be removed. Weeds growing on land are usually controlled mechanically or by herbicides; there is therefore less incentive to investigate uses for them as a means of control. Nevertheless, Savangikar & Joshi (1978) have extracted LP from *Parthenium hysterophorus*, which is becoming increasingly troublesome in India.

Soil from which weeds are regularly harvested will soon be depleted of nutrients, and there will be little further growth until these are replaced. Regular harvesting will usually alter the composition of the sward. Unless an area is manured and reseeded regularly, harvesting will eradicate some weed species. From this it follows that the successful exploitation of a wild plant, or 'weed', will necessitate the domestication and cultivation of it as a new crop; cogent reasons will therefore have to be given for cultivating a new species rather than a conventional one.

A few apparent defects of wild plants can be dismissed summarily. Though some grow vigorously, some seem unprepossessing simply because they are growing on poor soil. Even with good husbandry some, which may extract well in the laboratory, may grow slowly. A glance at pictures of the wild ancestors of our crop plants shows what improvements skilled selection can bring about. Uniformity in the time of germination is an important characteristic for which conventional crop plants have been selected. Wild plants tend to be much less uniform: this is an understandable aspect of evolution for it ensures that every individual plant of a species will not suffer should there be inclement weather throughout a region at about the time of germination. Uneven germination should not be considered a serious defect in an uncultivated species because uniformity could probably be introduced quickly by deliberate selection.

Many uncultivated plants are poisonous because of the presence of alkaloids and similar substances. This also is not a serious defect because alkaloids would be separated from the LP during the process of extraction (pp. 51, 56). Some species, however, are very strongly flavoured and it may be difficult to remove flavours from the LP. In my opinion it is a pity so much of the LP that is now being used is made from lucerne because most varieties have a strong flavour which is not easily removed from the LP. This should be borne in mind by anyone judging the acceptability of LP solely on the basis of experience with lucerne. When made from many other species, LP has little or no flavour. Nevertheless, LP from lucerne is indubitably edible (see Chapter 7); when made from a few other species it is not. An extreme example is *Leonotis taraxifolia* (called the duppy-pumpkin in Jamaica); it grows vigorously, extracts well and gives an unusable product. It is important therefore to assess the flavour of a carefully made preparation of LP at an early stage in the study of any new species.

The primary reason for suggesting that wild plants deserve investigation is that almost all our crop plants were chosen from among wild plants 5000 to 10 000 years ago. There is no reason to assume that our ancestors, at the time of what is often called the 'Neolithic Revolution', had unsurpassable skill in identifying potentially useful plants. Furthermore, as techniques of biochemical engineering evolve, it becomes possible to use hitherto useless plants (Pirie, 1962). Another reason, stressed by Hudson (1975, 1976) and by Wareing & Allen (1977) is the meteorological situation in Britain and several other countries. In spring, by the time there is abundant light and moisture, and day-time air temperature is adequate for growth, soil temperature is still too low for the roots of our conventional crops to be able to absorb nutrients. But low soil temperatures, at both the beginning and the end of the year, do not stop the growth of many weeds, e.g. dog's mercury (*Mercurialis perennis*), chickweed (*Stellaria media*), jack-by-the-hedge (*Sisymbrium alliaria*) and groundsel (*Senecio vulgaris*). For purely scientific reasons it would be interesting to know the basis of this physiological difference between crops and some weeds. It is not impossible that the useful ability to grow in cold soil could then be introduced into some conventional crops. At Rothamsted, nettles (*Urtica dioica*) yielded 612 kg of LP per ha; this is more than we got in one harvest from any other weed (Pirie, 1959c). Carlsson (1975) put nettles near the top in his survey of 71 species, but concluded that the Chenopodiums were probably the group most worthy of fuller investigation.

It would be unreasonable to devote more than a small fraction of the total effort put into research on LP to the study of unconventional species. Nevertheless, this study would be a useful piece of academic research. Besides the possibility that the period of active photosynthesis could be extended if more plant species were included in the cropping sequence, diversity, either of species growing together or grown in succession, offers many possible advantages. Suitably designed mixtures would contain species that grew when others were dormant, that exploited the full depth of soil for nutrients and water, and that were not all susceptible to the same pests and diseases.

4
Separation, purification, composition and fractionation

The manner in which LP will be used, the manner in which juice was extracted, and the species from which it was extracted, influence the procedures used for separation and purification. If LP is to be used as human food it should be as free from surface contaminants as possible. The crop should therefore be washed, before being pulped, to remove soil and dust from it. Crops from which LP is made will usually be harvested regularly, if perennial, and after a short period of growth if annual. They will therefore be less subject to attack by pests than conventional crops and there should be less need to use insecticides and herbicides. Use of these potentially toxic substances should be avoided until more is known about the extent to which they are removed by washing the leaves. The most convenient and thorough method of washing is to dump the crop into a tank of water, and drag it out continuously under a fine spray of water at a rate suited to the capacity of the equipment used for extraction. Wetting the crop in this way, although it increases the volume of juice handled, increases the percentage of protein extracted. A crop that will be washed must be harvested by a reaper of the old-fashioned type with a reciprocating blade. If it is harvested by a machine with flails or a rotor, and is then blown into a trailer, it will not only be contaminated with soil sucked up by the harvester, but will also be so bruised that much of the LP will be lost during the washing. There is less need to wash the crop when LP is being made as animal feed.

Juice made in an arrangement that separates pressing from pulping contains less fibre than juice made in any type of screw expeller because the fibre is not so heavily and assiduously rubbed across the barrel of the expeller (p. 134). Juice from any extraction unit that could be used on a large scale will initially contain an amount of fibre that would be unacceptable in human food. Unacceptability does not arise because the fibre is harmful: on the contrary, the ratio of fibre to protein will always be smaller than in conventional leafy vegetables, and there is a growing body of opinion that many people eat less fibre than is needed for effective gut behaviour (Nutrition Society, 1973). The main objection to incomplete removal of fibre is that the particles are clearly visible in dried LP and cause adverse comment. For many reasons (p. 79), LP should be used undried whenever possible. Fibre particles are then less obtrusive. Fibre is easily removed by passing leaf juice through any of the standard types of rubbed or shaken sieve. A simple rotary arrangement is illustrated in the IBP Handbook (Pirie, 1971*a*).

Preparation of unfractionated leaf protein

Most of the individual proteins in those leaf extracts that have been studied precipitate between pH 4 and 5: those proteins that do not precipitate in this range when purified, tend to precipitate along with the others when an unfractionated leaf extract is acidified. Acidification has been suggested as a method for separating LP in bulk although the curd produced is inconveniently soft and hydrophilic, and acidification does not kill many of the microorganisms that are inevitably present in leaf extracts. All the preparations of LP that have been used in extensive feeding trials were made by heat coagulation; Subba Rau & Singh (1970) report slower growth of rats fed on acid-coagulated than in those fed on heat-coagulated protein; acid-precipitated 'cytoplasm' protein is less digestible *in vivo* (Saunders *et al.*, 1973). Juices of the species of leaf from which bulk supplies of LP have up to now been made have pHs between 5.6 and 6.4. Extracts may, for several reasons, be more alkaline than this. Little protein would extract from naturally acid leaves unless alkali were added during the extraction, and if alkali is being added, it is essential to add enough to take the pH as far as 8. Otherwise pectin methyl esterase will liberate acid and cause the pH to drift back again (Holden, 1945). Addition of alkali increases the extractability of protein – it also increases the extractability of pectic substances. These coagulate along with the LP as harmless diluents. Alkali may be added to increase the stability of carotenoids (p. 93), and alkaline sulfite may be added to inhibit the oxidation of phenolic substances (p. 85).

Finally, some leaf extracts are naturally alkaline. Alkaline extracts from a few leaves coagulate satisfactorily on heating, others form a finely dispersed coagulum that is difficult to separate. That problem can be overcome by adjusting the pH. Manipulating the pH of an extract is troublesome, especially when LP production in a village is contemplated. As a rule, for the sake of simplicity, it will be best to forgo the advantages of alkalinity, either natural or induced.

All the protein in a leaf extract of suitable pH coagulates in a few seconds at 70 °C. Brief heating is sufficient; sudden heating is desirable because it produces a curd that is hard, granular and easy to manage on a filter (Morrison & Pirie, 1961); it also minimises enzymic changes. All the juice enzymes are not inactivated at 70 °C and they may cause changes in the protein during prolonged storage. With a crop such as lucerne, which contains more chlorophyllase than most crops, heating to 100 °C is advisable. Otherwise, the enzyme splits phytol from some of the chlorophyll and the resulting chlorophyllides readily lose magnesium to yield pheophorbides (Arkcoll & Holden, 1973; Holden, 1974) (p. 52). LP from a lucerne extract that had been heated slowly to 80 °C contained enough pheophorbide to photo-sensitise rats (Lohrey, Tapper & Hove, 1974; Tapper, Lohrey, Hove & Allison, 1975). That material also photosensitised pigs (Carr & Pearson, 1974), but a commercial product seems to have been prepared more carefully and caused little trouble (Carr & Pearson, 1976). This photosensitisation is analogous to hypericism caused by various plant pigments and would presumably (Darwin, 1868) affect only animals with white patches and kept in strong light, but it should be avoided if possible. Preparations heated to 100 °C absorb atmospheric oxygen more slowly than those heated to 70 °C (Shah, 1968), but there is so much nonenzymic oxidation when dry material is stored without protection that this difference is probably not significant. The only known disadvantage in heating extracts to 100 °C rather than 80 °C is the extra consumption of steam.

As soon as a leaf is pulped, autolysis begins; the rate at which protein is digested varies with the leaf species (Tracey, 1948; Singh, 1962; Batra *et al.*, 1976) and probably also with the age and nutritional status of the crop. For maximum recovery of protein, there should be no avoidable delay between pulping and heat coagulation:

some delay is advisable if nucleic acid in the LP is considered deleterious (p. 6). As often, incompatible requirements necessitate a compromise.

When working with a few litres of juice, sudden heating is easily managed by heating a pan of *water* to 80 °C and then, with vigorous stirring and continued heating, running in the cold juice at such a rate that the temperature in the pan never falls below 70 °C. The pan should have an overflow 5 to 8 cm up its side. The overflow should be 2 or 3 cm wide and about 10 cm long: the pan itself can conveniently be 15 to 30 cm wide. In large-scale work, steam injection is more convenient. A suitable unit for the purpose is a U-shaped, 3- to 5-cm (internal diameter) iron pipe with an outlet for curdled juice and an outlet to cope with surges on one side, and an inlet for the steam at the bottom and the inlet for cold juice at the top of the other leg. This simple arrangement (Pirie, 1957) has been complicated, without being improved, by so many people that it may be justifiable to reprint a drawing of it (Figure 4). Steam is turned on first, the unit will reach 90 °C in a few seconds, juice can then be run in at such a rate that it comes out at the required temperature. If that temperature

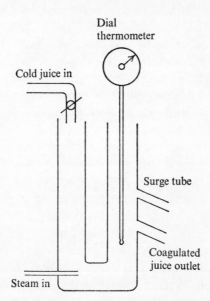

Figure 4. A convenient arrangement for coagulating leaf juice.

should be greater than 90 °C, the vertical arms of the U-shaped pipe should be extended so as to contain the inevitable surging. A thermostat valve on either the steam or juice inlet is a refinement that is hardly necessary because, if juice comes from a header tank in which the level is kept nearly constant, and if steam pressure is constant, the unit will run at constant temperature with minimal supervision. Every complication that is introduced seems to increase the risk that some underheated juice will get mixed with the coagulated juice: filtration will then be troublesome.

The air that is forced out of juice when it is heated makes most of the coagulum float; it can all be made to float by bubbling air (or preferably nitrogen) through the suspension. Alternatively, the suspension can be stirred gently until the coagulum sinks. These are refinements. If steam injection has been managed competently, the coagulum filters off so easily that little is gained by removing some of the 'whey' by decantation. Whatever course is adopted, it is advisable to separate the coagulum as quickly as possible because soluble phenolic substances in many leaf extracts combine with the protein if contact is prolonged; this diminishes the nitrogen content and digestibility of the LP (p. 84). When juice from less than 300 kg (wet weight) of leaf is being processed, the coagulum can conveniently be separated by pouring the coagulated juice into cloth 'stockings' about 18 cm wide (when flat) and 150 cm long. These are allowed to drip for as long as 'whey' runs out freely; they are then placed between corrugated mats and pressed at 0.3 to 0.5 kg per cm^2 (35 to 50 kPa) for a few hours. Pressure is most simply applied by means of a weighted lever. It is more important to have mats that allow the juice to escape freely than to press heavily. With larger quantities of leaf some form of belt press, or pump-fed filter press, will be needed. De Fremery, Bickoff & Kohler (1972), and Connell & Foxell (1976) studied the performance of several commercial models.

Dry leaf powders are used as flavouring agents and relishes in many parts of the world; there is therefore no universal dislike for 'leafy' flavours. However, during the initial phase of introducing LP as a human food, as much flavour as possible should be removed. Furthermore, some potential sources of LP, e.g. potato haulm, contain water-soluble materials that are slightly toxic; others, e.g. lucerne, contain growth depressants (Peterson, 1950); and all contain reducing sugars. The details of the Maillard reaction, in which lysine is made unavailable by combination with reducing sugar, have been studied by Finot & Mauron (1972). For all these reasons, as much soluble material as is practicable should be removed from LP.

Coagulum containing 40 % DM is easily made by pressing. If the residual water contains the amount of soluble material usually found in leaf extracts, about 7.5 % of the DM of the LP will be this extraneous soluble material. A less adequately pressed coagulum may contain 80 % of water; soluble material will then account for about 20 % of the DM. Similarly, when a well-pressed coagulum is resuspended in so much water that there is only 1 g or less of soluble material per litre, and pressed again till it contains 60 % of water, only 0.15 % of the final cake will be soluble. The recommended standard is < 1 % soluble material (Pirie, 1971*a*).

As already mentioned, the coagulum from a suddenly heated extract filters off easily; when resuspended in water at leaf pH, filtration is difficult. Filtration is easy if salt is added to the water or if the suspension is acidified to about pH 4. It is not essential to measure this pH. Different samples of press-cake are so similar that it is safe to assume that 200 ml of 2*N* acid will be sufficient for 1 kg (wet weight) of cake, and the pH is not critical. Besides facilitating filtration, acidification ensures freedom from alkaloids, and results in a product with the keeping qualities of cheese, pickles or sauerkraut. Unfortunately, acidification converts the chlorophylls completely to pheophytins through loss of magnesium. The dull green is less attractive to most people than the bright colour of neutral LP, and the risk of pheophorbide formation is increased. There is more loss of carotene from acid than neutral LP and the unsaturated fatty acids in it are more rapidly oxidised. If LP is regarded as something to be made on the farm, these defects of acid washing must probably be accepted. If it is regarded as an industrial product, it could be washed by centrifugation or, if washed by acidification and filtration, the washed, pressed and crumbled coagulum could be neutralised again by exposure to ammonia vapour.

Composition of unfractionated leaf protein

Components soluble in lipid solvents

The DM of LP, made in the manner described above, contains 20 to 30 % of material soluble in the usual lipid solvents. The amount of lipid is often underestimated because, although the material is largely soluble in ether or petroleum ether after it has been extracted, only part of it is released from combination with other components of LP by these solvents. The remainder can be extracted by a mixture of chloroform and methanol, or by a mixture of alcohol and ether in the presence of a little strong acid (Holden, 1952; Buchanan, 1969a, b; Byers, 1971b). The most noticeable, but nutritionally least important, component is the mixture of chlorophylls and their breakdown products. The importance of processing leaf extracts in such a manner that pheophorbide is not formed has already been stressed (p. 48). That hazard is easily avoided. Properly made LP, still containing the green pigments, should probably not be used as food by people with Refsum disease, a rare enzyme anomaly in the metabolism of phytanic acid derived from phytol (Herndon, Steinberg, Uhlendorf & Fales, 1969), nor by the equally small group with an intestinal flora that makes phylloerythrin (Clare, 1952).

Nutritionally and commercially the carotenoids are the most important components of the lipids in LP. Xanthophyll is the commercially valuable component because it makes chicken's legs yellow (Colker, Eskew & Aceto, 1948) and that enhances their appeal in many countries: trivial as this may seem, it is the main reason for commercial production of LP at present (1977). β carotene (provitamin A) is the nutritionally valuable component because vitamin A deficiency is widespread in those communities that no longer eat dark green leafy vegetables regularly (p. 92). There is little destruction of carotene during the separation and coagulation of leaf extracts; thereafter, especially if exposed to light, air and slightly acid conditions, it is destroyed (Walsh & Hauge, 1953; Arkcoll, 1973a; Arkcoll & Holden, 1973). If protected from these deleterious agents, little carotene is lost during many months storage. Some still remained in carefully stored lucerne meal after 27 years (Zscheile, 1973)! The β

(a)

Some substances that may be formed from Chlorophyll (a or b)

| Phytol removal by chlorophyllase | Magnesium removal by weak acid |

Chlorophyllide (bright green)   Pheophytin (dull green)

Magnesium removal
by weak acid

Pheophorbide (dull green)

Loss of methoxyl and reduction of a
vinyl group through microbial action

Phylloerythrin (brown)

Figure 5. (a) Chlorophyll breakdown chart. (b) The structure of chlorophyll A. Definite positions are not assigned to the double bonds in the main ring. The fluidity of the bonding in that structure is probably part of the reason for its effectiveness in photosynthesis.

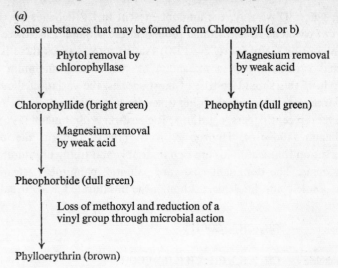

(b)

carotene content has not been measured on many samples of LP, but all contained 1 mg per g when freshly made and some contained

2 mg per g. Obviously, the amount present in LP depends on the ratio of carotene to extractable protein in the original leaf; there is an interesting claim (Savel'ev, 1970) that in some species there is a diurnal variation, with a maximum at sunrise and minimum at 13.00 h. If this should be true of most species, the variation should be borne in mind when choosing times for harvesting.

Several recent papers, e.g. Lima, Richardson & Stahmann (1965), Buchanan (1969a, b), Hudson & Karis (1973), confirm the old observation that leaf lipids are rich in doubly and trebly unsaturated fatty acids. The dominant acids are palmitic, palmitoleic, stearic, oleic, linoleic and linolenic. Although the unsaturated acids usually

Palmitic	$CH_3 (CH_2)_{14} COOH$
Palmitoleic	$CH_3 (CH_2)_5 CH : CH (CH_2)_7 COOH$
Stearic	$CH_3 (CH_2)_{16} COOH$
Oleic	$CH_3 (CH_2)_7 CH : CH (CH_2)_7 COOH$
Linoleic	$CH_3 (CH_2)_4 CH : CH CH_2 CH : CH (CH_2)_7 COOH$
Linolenic	$CH_3 CH_2 CH : CH CH_2 CH : CH CH_2 CH : CH (CH_2)_7 COOH$

Figure 6. The dominant fatty acids in leaf protein.

account for half the total fatty acid there are marked apparent differences in their ratios in different species. It is at present not clear what importance should be attached to these differences because the fatty acid pattern of leaves changes with age (Eichenberger & Grob, 1965; Kannangara & Stumpf, 1972; Hudson & Karis, 1974) and there is no reason to think that the crops used in the earlier experiments were of similar physiological age. This point deserves thorough investigation because of the nutritional importance of some unsaturated fatty acids (Elliott & Knight, 1972), and because their presence in LP complicates preservation and storage. Hudson & Karis (1974) point out that some leaf crops produce more lipid per ha in a shorter time than most oilseed crops.

Carbohydrates

The amount of starch in LP depends on the species used and the weather at the time of harvest. There is so much starch in extracts from some leaves, notably peas taken on a sunny day, that grains collect as a white sediment if the extract is allowed to stand. Unless the starch is centrifuged out of the extract before coagulation, 5 % (or even more) will be present in the LP. Even when no starch grains are visible in the extract, there is usually 5 to 10 % of carbohydrate in LP. By graded hydrolysis followed by paper chromatography, Festenstein (1976) showed that this carbohydrate differs in composition from the carbohydrate mixture in the 'whey' that is filtered from the coagulum. That is to say, the presence of this carbohydrate is not a consequence of the LP being inadequately washed. When alkali is added during the extraction, the carbohydrate content is increased because of the presence of pectic substances. Nucleic acid supplies a little of the carbohydrate in most preparations (p. 17).

Miscellaneous components

If LP has been thoroughly washed and pressed it will contain less than 1 % of water-soluble material. Consequently it will be nearly free from B vitamins and other such components of the leaf: their presence (Bray, 1976) is evidence of inadequate washing. LP will, however, contain useful amounts of vitamins E and K (p. 94).

The ash content varies. It is seldom less than 3 %, and may be as great as 8 %. Dust from the surface of the leaves is the source of much of it, but some of the grasses contain silica which is partly extracted along with the protein. Silica appears to be harmless: there is even the suggestion (Schwarz, 1977) that it is beneficial. Metallic cations, such as Ca^{2+} and K^+, are to a large extent removed during the washing at pH 4; that is near the isoelectric point of most of the proteins that make up LP. An IBP Technical Group (Pirie, 1971*a*) suggested that when LP is to be used as human food it should contain < 3 % ash and < 1 % acid-insoluble ash. Later experience suggests that it may not always be possible to meet that standard. Nevertheless, material containing more ash than that should be carefully examined to see whether the ash is an unavoidable component of LP made from the species that is being used, or whether it is present because of some deficiency in the technique of preparation.

Plants grown on some soils absorb deleterious metals, such as lead and zinc, and translocate them to their leaves. When leafy vegetables are eaten in the conventional way, little attention is paid to the presence of lead in them though amounts as large as 50 mg per kg of DM have been recorded. Although lead is known to be more abundant in the chloroplasts than in other parts of the leaf (Holl & Hampp, 1975) much of the lead in leaves may be the result of surface contamination with earth, and the plants may have come from roadside sites where they are fouled by lead from motor car exhaust. There is apparently no legal restriction on the amount of lead that may be present in a 'natural' food; restrictions apply as soon as a food is processed. This may restrict the sites where crops for LP production can be grown unless some method, such as coagulation in the presence of a chelating agent, can be found that limits the amount of lead accompanying the LP.

The presence of alkaloids in a leaf species is unlikely to be an obstacle to its use as a source of LP because all the alkaloids are soluble on the acid side of neutrality; they will probably be removed when LP is washed at pH 4. This expectation must obviously be verified for each species. Other components, such as the saponins in lucerne and the estrogens (p. 96) in some clovers (Glencross, Festenstein & King, 1972), may present more problems. Fortunately, there are varieties of most cultivated species that contain only small concentrations of these substances. Until the risk has been shown to be negligible, any crop suffering from a heavy fungus attack should be rejected lest a mycotoxin should be carried through into the LP. This last risk is not peculiar to LP: many edible seeds and tubers, especially after storage, may be contaminated with mycotoxins.

After extraction with lipid solvents, and after allowing for the presence of carbohydrate and ash, bulk preparations of LP invariably contain less nitrogen than is usually found in proteins. Phenolic compounds and other tanning agents are present in all species of leaf; in some the quantity is large. Besides interfering with protein extraction (p. 83), it is reasonable to assume that some of these substances accompany what protein does extract, and thus con-

taminate the final LP. Jennings, Pusztai, Synge & Watt, (1968) measured the absorption spectrum of bean (*Vicia faba*) leaf protein, compared it with the spectrum calculated from its tyrosine and tryptophan
content, and showed that the difference between the spectra resembled
the typical spectrum of phenolic material. The extreme values for the
phenolics in LP from nine species were 1.4 and 2.2 % (Subba Rau
et al., 1972); phenolics were therefore not major components of any
of these LPs, but those in which the ratio of phenolics to nitrogen was
small were the preparations with the best nutritional value (p. 86).

When systematic work starts on the selection of species best suited
for large-scale LP production, and on the optimum time of harvest,
a close watch will have to be kept on the phenolics. It is already well
known that they tend to increase with maturity or stress (e.g. Milic,
1972; Wong, 1973). When studying the extent to which they impede
protein extraction, and damage what protein is extracted, some of
the methods that are used in the laboratory to promote the extraction
of enzymes or viruses should be tried. For example: air could be
excluded during extraction (Cohen *et al.*, 1956; Pirie, 1961), a reducing environment could be maintained with —SH compounds (Hageman & Waygood, 1959; Anderson & Rowan, 1967), or substances
such as nicotine could be added to compete with protein for conjugation with the phenolics (Thung & Want, 1951; Thresh, 1956). More
recently, polyvinylpyrrolidone has been used in many operations in
which it is desirable to sequestrate tanning agents. When added to
grass or lucerne extracts before heat coagulation the amount of
phenolic material in the LP was halved (Fafunso & Byers, 1977).
Methods such as these could probably not be used in large-scale work,
but experiments along these lines show what the potential is and
which factors prevent it from being fully exploited. In large-scale
work it is already possible to make quantitative and qualitative improvements by adding sulfite during the extraction; it may be possible
to devise methods for removing phenolics while the crop is being
washed before being pulped.

Amino acids

There is no need to survey the early literature on the amino acid
composition of LP. These analyses showed that LP should have good

nutritional value; they were therefore the justification for all the work that has been done on methods of extraction, but they have been superseded. Byers (1971*b*) described the precautions that must be taken to get trustworthy results on material containing considerable amounts of carbohydrate and lipid. The only addition that I would make to her recommendations is that all analyses should be done on at least two preparations, made from the same species in the same way. With automated equipment, the amount of work involved in an analysis is so small that editors should not permit journal space to be wasted on analyses done less thoroughly. Furthermore, trustworthy amino acid analyses on LP from many species, and from plants of differing age, show such uniformity (e.g. Bryant & Fowden, 1959; Pleshkov & Fowden, 1959; Gerloff, Lima & Stahmann, 1965; Byers, 1971*a*, *b*; Fafunso & Byers, 1977) that there is little need to publish further similar results.

Uniformity in amino acid composition is, at first sight, surprising because there are great differences in the specific enzyme activities of leaves from different species, or differing in age. However, apart from ribulose-diphosphate carboxylase, each individual enzyme probably accounts for such a small fraction of the total LP that, even although they have different amino acid compositions, the average composition is little affected by changes in enzyme ratios. These considerations lead to the conclusion that one of the less necessary pieces of equipment to install in an institute where work on LP is starting is an amino acid analyser. If a species should be found that extracts well and is promising from an agronomic standpoint, perhaps from a family from which LP has not hitherto been made, it is likely that some institute already equipped with an analyser will be willing to analyse one or two samples. In the unlikely event that the LP has an unusual composition, more detailed cooperation can be organised.

Table 2 (from Byers, 1971*a*) illustrates the reason for expecting uniformity in amino acid composition. The differences between the three species, and between the unfractionated LP and the 'chloroplast' and 'cytoplasm' fractions (p. 62) made from it, hardly lie outside the range of unavoidable uncertainty in amino acid analysis. This uncertainty arises not only from errors in measuring the areas of the peaks on the elution profile given by an automatic analyser, but also from the fact that if less than 100 % of the nitrogen in the sample is accounted for in the amino acids recovered there is no guarantee that the losses are distributed among the amino acids in the same way in all samples. Nevertheless, the small differences in amino acid composition are probably real. This was demonstrated by an elaborate statistical study (Byers, 1971*b*). Essentially, the difference between the largest and smallest content of each amino acid was found for the whole group of 39 samples. Then, the amino acid contents of each sample were compared in turn with the amino acid contents of each of the other 39 minus 1 samples. The criterion of comparison was the extent to which the difference in each amino acid content in each pair was less than the difference for that amino acid in the whole group of 39. Because cyst(e)ine and tryptophan are partly or wholly destroyed during acid hydrolysis, this was done for only 16 amino acids. It nevertheless produced a formidable collection of figures. The next process may metaphorically be described as arranging the figures in 39 minus 1 dimensional space and instructing a computer

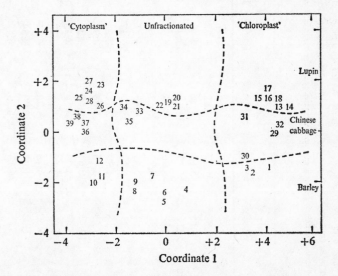

Figure 7. Statistical analysis of the amino acid composition of 39 preparations of leaf protein. By a technique outlined in the text, the analyses were arranged by a computer in a manner designed to demonstrate the reality (if any) of differences in composition between preparations of different types. The numbers within the figure have no quantitative significance, they merely identify preparations.

Table 2. Amino acid composition (expressed as weight of amino acid (g) per 100 g of recovered amino acids) of three types of leaf protein preparation from three species. Ammonia, cyst(e)ine and tryptophan are excluded. (From Byers, 1971a)

Amino acid	'Chloroplast'			Unfractionated			'Cytoplasm'		
	Barley	Lupin	Chinese cabbage	Barley	Lupin	Chinese cabbage	Barley	Lupin	Chinese cabbage
Aspartic acid	9.75	10.10	9.86	9.57	10.22	10.01	9.62	10.01	10.05
Threonine	4.82	4.97	4.87	5.07	5.01	5.22	5.41	5.04	5.43
Serine	4.85	5.15	5.24	4.40	4.68	4.50	4.10	4.13	4.12
Glutamic acid	11.00	11.35	11.28	11.41	11.88	11.91	11.94	12.15	12.21
Proline	4.88	5.06	4.98	4.68	4.79	4.72	4.62	4.79	4.46
Glycine	6.12	5.97	6.53	5.64	5.69	5.35	5.38	5.32	5.29
Alanine	7.05	6.40	6.45	6.71	6.21	6.10	6.52	5.99	5.91
Valine	6.16	6.10	5.64	6.37	6.27	6.06	6.50	6.32	6.17
Methionine	2.28	1.90	2.11	2.24	1.70	1.94	2.39	1.76	2.13
Isoleucine	5.25	5.76	5.03	4.95	4.93	4.62	4.74	4.42	4.38
Leucine	10.43	10.68	10.40	9.33	9.75	9.29	8.42	9.21	8.79
Tyrosine	4.49	4.20	4.19	4.50	4.61	4.71	4.92	5.56	5.07
Phenylalanine	6.97	7.16	6.85	6.22	6.24	6.24	5.84	5.82	5.87
Lysine	5.60	4.78	5.23	6.61	6.60	7.08	7.06	7.30	7.23
Histidine	1.82	1.91	2.00	2.34	2.31	2.42	2.66	2.82	2.65
Arginine	6.29	6.13	6.33	6.89	6.35	6.46	7.01	6.67	6.89

to search for the plane in that space that contained the largest differences in amino acid composition. Figure 7 shows the 39 samples plotted in accordance with their two coordinates in that plane. Although the three types of preparation, from three species, are not separated completely, it is clear that there is some grouping into nine categories.

Gerloff *et al.* (1965) included tryptophan in their set of analyses. If attention is restricted to preparations containing more than 60 % protein, all the values are in the range 1.3 to 2.2 %; when results are expressed as in Table 2, that comes to 2 to 3 g of amino acid per 100 g of total amino acids. The scatter is not unexpected; tryptophan determinations are notoriously unsatisfactory, and some heavily contaminated samples of LP gave suspiciously large values. Cyst(e)-ine is also not included in the table because of uncertainty about the extent to which it is destroyed during hydrolysis in the presence of phenolic substances (Fafunso & Byers, 1977 and p. 87) and of substances that form humin. Destruction is complete if air is not excluded. Even with careful exclusion of air, Byers' (1971*b*) values varied from 0.69 to 3.04 g of cyst(e)ine per 100 g of recovered amino acids in the nine types of preparation – a range very much larger than that found with any other amino acid. Well-washed LP that has been thoroughly extracted with lipid solvents is not known to contain any sulfur in forms other than cyst(e)ine and methionine. An upper limit for the cyst(e)ine content can therefore be deduced from the difference between the total sulfur and the sulfur contained in methionine (Subba Rau *et al.*, 1972; Byers, 1975). Byers analysed 32 preparations; by direct analysis the percentage of cyst(e)ine in the true protein component of LP from lucerne was 0.82 to 2.15, and by calculation 2.35 to 2.78. With lupin (*Lupinus albus*) LP the corresponding values were 'trace' to 2.79 and 2.00 to 2.67. Obviously, the validity of the method depends on the validity of the assumption that sulfur is never present in LP in a third form.

Fractionation of leaf protein

As a prelude to making enzyme or virus preparations from leaf extracts, chloroplasts and their fragments (which distressed Osborne so much (p. 4)) have, for half a century, been coagulated by freezing, addition of calcium salts, judicious addition of solvents such as ethanol, or of weak acids. It is unlikely that these methods could be used in large-scale work either because of expense or of the need for very close adherence to precise conditions. The methods that are most likely to be applicable are heating, the addition of water-immiscible solvents or ionised polymers, and high-speed centrifuging. The coagulum is loosely called 'chloroplast' protein because it carries with it the chlorophylls and their green breakdown products; the protein that remains in solution is, equally loosely, called 'cytoplasm' protein. The 'chloroplast' fraction carries out with it components of such structures as nuclei, mitochondria and ribosomes; when made from leaf extracts that have been inadequately strained, or made from inadequately washed leaves, it also carries any fibre and dust present in the extract. For all these reasons, 'chloroplast' protein usually contains 6 to 8 % nitrogen, unfractionated protein contains 9 to 11 %, and 'cytoplasm' protein may contain 16 %. 'Chloroplast' protein is less digestible (p. 76) both *in vitro* and *in vivo*; the quality of the remaining protein is therefore improved if it is removed. These are real merits and it may be that when LP is produced industrially it will be worthwhile separating the nearly colourless 'cytoplasm' fraction because it could be presented as a food in more ways than the 'chloroplast' fraction. But the separation is not simple enough to be regarded as a farm operation. Whatever method is used for fractionating leaf extracts, the precise conditions that give complete separation of the 'chloroplast' fraction with minimum loss of 'cytoplasm' protein probably depend not only on the leaf species, but also on its age, fertiliser treatment and the weather to which it was exposed. It is unlikely that fractionation will be satisfactory if a prearranged schedule is followed; instead, some preliminary laboratory examination of each type of extract will probably be necessary.

Most of the protein in a leaf is present initially in the chloroplasts. Even the refined methods used to separate them in the laboratory release some of their contents. The rougher methods used in bulk extraction probably cause still more damage. One of the factors that distinguishes species classified as good sources of LP is probably chloroplast fragility, and one of the advantages of pulping with added water may be that osmotic shock when chloroplasts are released into dilute suspension disrupts them more thoroughly and so allows more of the total protein of the leaf to appear in the 'cytoplasm' fraction. Nevertheless, it is seldom possible so to arrange conditions that less than two-thirds of the protein appears in the 'chloroplast' fraction, and this fraction contains the carotene. It will be a pity if it is all used as animal feed in countries where there is a shortage of vitamin A (p. 92) in human diets.

Fractionation by heating

Rouelle's (1773) method of heating in two stages (p. 139) is still the most widely used fractionation procedure. Heat coagulation is so rapid that even momentary local overheating must be avoided. A few 100 ml of juice can be swirled in a flask while being heated directly. When working on a large scale it is better to maintain one surface of a heat-exchanger at the required temperature and to let the juice flow over the other surface while it is being continuously scraped (Pirie, 1964*b*). The juice must then be centrifuged because the curd is too soft and gelatinous to be collected on a filter. Byers (1967*b*) found that 61 to 64 % of the protein from lupin was in the 'chloroplast' fraction made in this way, 80 % of the protein from rape and 92 % from nasturtium (*Tropaeolum majus*). These measurements were made at the natural pH of the leaf extracts, and at 43 °C with lupin but 53 °C with the other two. Working at 53 °C throughout, Lexander *et al.* (1970) adjusted the pH of extracts from 14 species to values between 4.5 and 6, and found, as others have done, that the proportion of the protein separating in the 'chloroplast' fraction increased as the pH diminished. With extracts from frozen sunflower leaves, all the protein was in that fraction even at pH 6; the smallest amount of protein was in that fraction with nettles, lucerne and *Atriplex hortensis*. This work was extended by Carlsson (1975) who heated extracts from 41 species to 53 °C at their natural pHs. The more alkaline leaves produced the least 'chloroplast' coagulum: of those leaves that extracted well, the ones giving the smallest amounts of 'chloroplast' protein were beet (32 % at pH 7.1), spinach (43 % at pH 6.5) and *Atriplex latifolium* (50 % at pH 6.6). These figures are quoted to show how large the differences can be between extracts from different species, or between extracts from the same species when the conditions of heating and centrifuging differ.

'Cytoplasm' protein coagulates when juice from which 'chloroplast' protein has been removed is heated to 80 or 100 °C. After pilot-plant studies (de Fremery *et al.*, 1973; Edwards *et al.*, 1975), the process was patented (Bickoff & Kohler, 1974) in spite of the long period during which it has been in common use. To avoid loss of coagulable protein, caused by the action of proteolytic enzymes in the leaf extract, the extract must be heated quickly to about 60 °C, held at that temperature for only a few seconds and then cooled quickly before being centrifuged. In the arrangement used at Rothamsted (Pirie, 1964*a*), the outflow from the scraped heat-exchanger passed through a metal tube cooled with ice. The supernatant fluid then has to be reheated. It may be possible to manage these manipulations economically, but it would not be prudent to assume this. The first phase of the separation has, however, considerable academic interest. For example, Jones & Mangan (1976), using lucerne extracts to which inhibitors of oxidative processes had been added, removed the 'chloroplast' fraction and then separated a homogeneous 'cytoplasm' fraction with the physical properties of ribulose-diphosphate carboxylase.

The preparation of partly dried fibre, that can be economically conserved as winter fodder by complete drying, and that is not greatly depleted of protein, is likely to be one of the objectives of fodder fractionation (p. 113). This can be achieved by heating the leaf to destroy its osmotic control and then pressing without rubbing or disintegration. The technique was tried 40 years ago (cf. Pirie, 1977*a*) and is being investigated again (Mathismoen, 1974; Gastineau, 1976). In these trials almost all the protein was coagulated *in situ*. By heating to a lower temperature, perhaps 45 to 55 °C, it should be possible to coagulate the 'chloroplast' protein while leaving the 'cytoplasm' protein extractable.

Centrifugal separation

The feasibility of separating 'chloroplast' protein centrifugally on a commercial scale depends entirely on choosing crops with chloroplasts that do not break up into very small fragments, and on the availability of robust centrifuges that can be run economically at the required speed. There is nothing novel in the principle. After 1940, there were major improvements in centrifuge technology that were immediately made use of in plant virus research. There is reasonable agreement (e.g. Bawden & Pirie, 1938; Lugg, 1939; Pirie, 1950, 1955; Pierpoint, 1959; Chayen *et al.*, 1961; Wilson & Tilley, 1965; Byers, 1971*b*; de Fremery *et al.*, 1973) that chloroplasts and their fragments sediment at 8000 to 50 000 times gravity depending on the duration of centrifuging and the depth of fluid through which a particle has to travel to be effectively sedimented. In a centrifuge with swinging buckets, this depth equals the length of the tube; in an angle-head centrifuge, it is not much greater than the width of the tube; in a continuous-flow centrifuge, it is a millimeter or less. To sediment all the green material it is usually necessary to centrifuge at 100 000 times gravity. The material removed by high-speed centrifuging is similar to, but not identical with, the material removed by heating to 50 or 60 °C; the supernatant fluid after centrifuging usually still contains some protein that coagulates when heated to 60 °C. If it should prove to be feasible on a commercial scale, centrifuging will be a valuable pretreatment in making good-quality LP from some species. Thus LP made from potato leaf juice that had been centrifuged at 25 000 times gravity contained 11 to 12 % nitrogen whereas without that treatment it contained only 9 to 10 % (Carruthers & Pirie, 1975).

Fractionation with solvents

Material similar to the 'chloroplast' protein that coagulates on gentle heating is precipitated more easily than the remainder of the protein by solvents such as methanol, ethanol and isopropanol. But 10 to 20 % of the solvent has to be added and recovery of this would be uneconomic. Tsuchihashi (1923) coagulated the cell walls in a suspension of hemolysed erythrocytes by shaking with chloroform. This process was improved by adding water-immiscible solvents, such as butyl alcohol, to the chloroform and, before the advent of gel-filtration, it was used extensively to remove almost all the protein from solutions of other macromolecules (e.g. Heidelberger, Kendall & Scherp, 1936). By suitable choice of solvents and duration of shaking, the method can be used to coagulate some of the component proteins in a mixture while leaving others in solution. As might be expected in view of the extent to which they resemble erythrocytes, chloroplasts and their fragments are readily coagulated by many water-immiscible solvents. Slade, Branscombe & McGowan (1945) used amyl alcohol; Allison (quoted in Hove & Bailey, 1975) used *n*-butanol. Some of the halogenated hydrocarbons have been used in similar processes; they may be preferable in the laboratory, but they are more expensive. Furthermore, if the 'whey' that runs away from the heat-coagulated 'cytoplasm' protein is to be used as a culture medium for microorganisms (p. 120), there are advantages in using a solvent that can be metabolised by microorganisms and that acts as a coagulant at a concentration small enough not to inhibit their growth.

Miscellaneous methods of fractionation

Chibnall's (1939) method of removing the larger particles, including 'chloroplast' protein, by filtering leaf extracts through a 3-cm-thick pad of paper pulp, has occasionally been used by others (e.g. Lugg, 1939; Davies, Evans & Parr, 1952; Yemm & Folkes, 1953; Chibnall, Rees & Lugg, 1963). In the original form, it is satisfactory for making 'cytoplasm' protein, but the 'chloroplast' fraction would be difficult to recover in a usable form. Technical developments in the production of smooth-surfaced membranes of graded porosity, that can be scraped clear of 'chloroplast' protein when they get clogged, may make such a method practicable on a commercial scale.

The 'cytoplasm' protein in solutions that have been freed from 'chloroplast' protein by any of the methods described above can be concentrated five- to tenfold by ultrafiltration (Tragardh, 1974; Knuckles, de Fremery, Bickoff & Kohler, 1975). Much of the brown pigment and the soluble salts are removed so that a pale product containing $> 93 \%$ protein can be made by precipitation with alcohol, heat or acid. If the temperature is kept at or about 0 °C during these operations, acid-precipitated material will redissolve at neutrality or at pH < 3. Ostrowski (in Wallace, 1975) suggests that the yield of LP is greater when it is made by ultrafiltration than by heating. The wording is not clear: it may be that the yield of 'cytoplasm' protein is meant. That is credible, but the product would probably still contain 'whey' polysaccharide. Protein has not as yet been fractionated in this way on a commercial scale. Because of its solubility, the product would fit more smoothly than heat-coagulated material into the conventional operations of the food industry, but it would cost very much more to produce.

Before high-speed centrifuges were standard pieces of laboratory equipment, clarified leaf extracts were often made for serological and other studies by adding enough alkaline phosphate to precipitate the Ca^{2+} ions in the extract. The calcium phosphate precipitate carries out with it all the 'chloroplast' protein: it also carries out the nucleic acid (Pirie, 1974). This procedure deserves reinvestigation. Brief reference is sometimes made (e.g. Bray, 1977) to the coagulation of 'chloroplast' protein by gums and resins containing polar groups. If the necessary amount of coagulating agent is small enough, processes such as this could be useful in commercial practice. Concentrated extracts from some plant species, e.g. various tobacco (*Nicotiana*) varieties, deposit crystals containing ribulose-diphosphate carboxylase when dialysed (Chan, Sakano, Singh & Wildman, 1972). In spite of the technical difficulty of large-scale dialysis, this process has been suggested as a means of making edible protein. If this suggestion is being taken seriously, leaf extracts cleared of 'chloroplast' protein by the addition of phosphate would seem to be the most suitable starting product.

5
Preservation, storage and modification

Unfractionated preparations of LP have often had different nutritive values in spite of having comparable nitrogen contents. These differences are probably more often the consequence of differences in the techniques used in preparation and preservation, than of differences in the ages or species of the leaves. Because all the reactions that take place during preservation are likely to be harmful, LP, like many other foods, should be used in the fresh state whenever possible. But preservation will often be necessary if LP is to be used at a distant place, or in a season when crops do not grow. Similarly, when unfractionated juice is used as chicken or pig feed, preservation may be needed because there will be fluctuations in the supply in normal farm conditions.

Preservation of leaf juice

Leaf extracts vary in proteolytic activity (p. 48), but most of the protein in all of them hydrolyses in a few days at room temperature. This does not necessarily diminish the feeding value because, although protein nitrogen is converted to nonprotein nitrogen (if these are measured by conventional methods), the amino acids are still there and are presumably as valuable nutritionally as they were when in the original protein. Leaf extract is, however, an excellent microbial culture medium and it stinks within a few days. Extracts that have been heated to > 80 °C are partly sterilised and will then keep for a week – the advantages of some separation of protein coagulum from the 'whey', when nonruminants are being fed, are discussed in a later chapter (p. 132). For more prolonged preservation it is necessary to add a strong acid, e.g. formic, hydrochloric, or phosphoric, till the extract is at pH < 3, or to add sulfite as in the well-known method of preserving fruit. These techniques are discussed briefly in recent annual reports from the National Institute for Research in Dairying (Shinfield, Reading), the Rowett Research Institute (Aberdeen) and the Dunsinea Research Centre (Castleknock, Eire), also by Houseman & Connell (1976) and in several papers given at a symposium organised by the British Grassland Society (Wilkins, 1977).

Preservation of moist leaf protein

When LP is being washed to remove flavour, salts and carbohydrates that would damage the protein by Maillard (or 'browning') reactions (p. 51), it is acidified to about pH 4 to facilitate filtration. This brings it into the pH range of cheese, pickles, sauerkraut, or bottled fruit, and it has the keeping qualities of these foods. For prolonged storage, it can be canned. One attempt was made to preserve the moist cake by exposure to 2 Mrads from a cobalt source, but material that had been treated in this way had an unusual flavour that would probably be unacceptable.

Simpler methods for preserving the moist cake become more satisfactory the more completely water has been pressed out. When pressing, it is more important to arrange conditions in such a manner that the expressed juice can flow away easily, than to apply heavy and prolonged pressure (p. 50). Fully pressed material contains < 60 % water. If such material is mixed with one-seventh of its weight of salt, rammed into a jar, and protected from access of air in the ways familiar in jam-making, it keeps well. Each g of LP DM is then accompanied by about 0.25 g of salt. This is not an unreasonable amount of salt in a food provided the other components of the day's diet are not also heavily salted. The salt can, however, be partly removed before cooking by soaking the mass in water and letting the LP settle. If the final food is to be sweet, sugar can be used in the same way (p. 109); press-cake keeps well if it contains approximately equal weights of LP DM and sugar. Subba Rao, Singh & Prasanappa (1967) preserved acidified cake for several months in the presence of 2 % residual acetic acid and 0.2 % orange peel oil. They examined the microbial population of the material in some detail. Arkcoll (1973a), who preserved LP from seven species for six months with 1 % acetic acid, found that the common soil fungus *Mucor racemosus* was the usual contaminant. This initial contamination is unavoidable. To avoid further contamination of the juice, the techniques of dairying are needed. The production of LP has much in common with the production of milk: there is a dirty side to the work and a clean side, and the two must be kept separate.

Foods are often preserved by encouraging a desirable growth instead of by preventing all microbial growth. Slade *et al.* (1939) suggested inoculating LP so as to make 'cheese'. This is a possibility because many highly esteemed cheeses have, for the uninitiated, peculiar flavours; but a liking for exotic cheeses is not quickly established. In cooperation with the National Institute for Research in Dairying, acceptable samples of ordinary cheese containing 5 to 10 % of LP were made. If LP were being preserved in this way, it would be worth considering the addition of *Streptomyces* to the inoculum so as to produce vitamin B 12 – this is often lacking in vegetarian diets.

Methods of drying leaf protein

Freeze-drying is the least harmful and the most expensive method of drying LP; it would be used only for the preparation of analytical samples or experimental diets. Fully pressed LP gives a rather hard and granular product when freeze-dried; a much more attractive product is made if the cake is moistened so as to make a paste containing 70 to 75 % water. The more quickly it is then frozen, the better the texture of the product; it is lumpy when large ice crystals are allowed to grow. By evaporative cooling in a suitable unit (Pirie, 1964c) a layer of paste 1 cm thick can be frozen solid in 1 to 2 min. Freeze-dried material appears to keep indefinitely if stored in an air-tight container but, because of its enormous surface, the unsaturated fats in it absorb oxygen rapidly if it is exposed to air. Exposure to light increases the rate of oxygen uptake. Lea & Parr (1961) concluded that, in spite of the heat treatment to which LP had been exposed, enzyme reactions were responsible for part of the oxygen uptake. It is inhibited by the usual antioxidants. LP made from leaf juice that had stood for 2 h at room temperature before coagulation oxidised faster than LP coagulated quickly; Arkcoll (1973a) suggested that this was the result of destruction of the natural antioxidant tocopherol (vitamin E), normally present in leaf juice. Hudson & Warwick (1977) suggest that more powerful antoxidants are also present in LP.

When LP is dried completely in a current of air, either at room temperature or in an oven, the product is dark and gritty. If press-cake is mixed with salt before drying, in air at room temperature, the product is more attractive but, if salt is being added, it might as well be added as a preservative to moist cake. LP will finally be mixed with some form of flour or meal before use as human or animal food. The processes that lead to darkening and hardening are completely prevented if these substances are used as 'extenders', that is to say, if drying is postponed until most of the components of the diet have been mixed intimately with moist LP. Many components of practical diets contain 10 to 12 % water when in equilibrium with an atmosphere of average humidity, and can be parched in an oven without loss of nutritive value. Duckworth & Woodham (1961), and Foot (1974), improved the performance of 'extenders' by parching them before mixing in the moist LP. Dry LP without the unattractive appearance typical of material dried directly in air can also be made by drying in a current of air in a tumble drier until the water content is only 20 to 30 % (this point is easily judged by feel): the material is then ground finely before drying is completed (Arkcoll, 1969). The performance of some commercial drying units was examined by de Fremery *et al.* (1972).

The extent to which LP is damaged when dried in an oven depends not only on the temperature of the air to which it is exposed, but also on the extent to which the LP had been washed to remove sugars, and on the rate at which water is being removed. The more rapid the evaporation, the greater the temperature difference between the LP and the air, and the shorter the time for which LP is moist when it is being heated. Buchanan (1969b) found that digestibility *in vitro* diminished more when LP was heated when it contained 7 % water than when it contained 3 %: from this it follows that a small amount of crumbled LP, dried quickly in a thin layer in a current of air, is likely to be damaged less than a larger amount dried in the form of lumps in a closed oven, even though the latter is being dried at a lower temperature. This is probably the explanation of the apparent conflict between the observation (Duckworth & Woodham, 1961) that LP was damaged by drying at temperatures above 82 °C, whereas Subba Rau & Singh (1970) found no damage during drying at 100 °C in a fast current of air. The precise conditions of drying were not stated by Munshi, Wagle & Thapar (1974) who found, using the growth of rats and the amount of xanthine oxidase in their livers as criteria, that cowpea LP was damaged more by drying at 80 °C than at 60 °C. The consequences of spray drying, or roller drying, incompletely washed LP or unfractionated leaf juice are hard to interpret. Although Cowlishaw, Eyles, Raymond & Tilley (1956*a*, *b*) had already demonstrated the poor nutritional value of dried whole juice for chicks, this technique was advocated by Hartman, Akeson & Stahmann (1967). The inadvisability of using the technique was stressed by Subba Rau, Mahadeviah & Singh (1969) and Pirie (1971*a*): many components of 'whey' that would be included in the LP have doubtful nutritional value for nonruminants and some of them are toxic.

Drying by solvent extraction

Solvents that are miscible with water, e.g. acetone and the lower alcohols, can be used for removing water and, at the same time, removing some of the lipids and chlorophyll and its breakdown products. This policy is advocated by those who make the curious assumption that a food should never be green. Although removal of the green colour is marginally advantageous, removal of the lipids is detrimental. They contribute nearly as much to the energy content of an average sample of LP as the protein contributes, and the carotene is a valuable component that would be lost (p. 92). A further disadvantage of solvent extraction is that removal of the solvent at the end of the process is nearly as difficult as the direct removal of water would have been. When acetone is used, it is almost impossible to remove the smell of mesitylene and its derivatives completely.

Solvent extraction has, on the other hand, some compensating advantages. The removal of unsaturated lipids, even if it is incomplete, makes it less necessary to exclude air from material that will be stored, and thorough solvent extraction can restore digestibility to LP that has been damaged by prolonged heating when moist

(Shah, Riaz-ud-Din & Salam, 1967; Buchanan, 1969*a*, *b*). At room temperature, the lipids become a little less easily extractable from moist LP within a few days: presumably by a process comparable to the tanning of 'chamois' leather. Buchanan (1969*b*) measured the progress of fixation at 60 °C in the presence of 1 to 2.5 % water. Because of this fixation, and because of the risk of microbial attack, solvent extraction, if it is decided on, should start within a few days of pressing. The lipids in dry soy LP (*Glycine max*) do not seem to become less extractable after storing at room temperature (Betschart & Kinsella, 1974*a*). Mokady & Zimmermann (1966) compared products made by extraction with methanol followed by chloroform, acetone followed by petrol ether, and boiling toluene, used first to distil off the water azeotropically, and then as a lipid solvent; the first system gave the most attractive product. There is general agreement (e.g. Huang *et al.*, 1971; Bray, Humphries & Ineritei, 1978) that polar solvents extract lipids most effectively from LP.

Modification

After being made insoluble at neutrality by heat coagulation, LP is still partly soluble at pH 9 or above. Exposure to the level of alkalinity that is needed to get much of the protein into solution is well known to damage other proteins by racemising some amino acids, by converting alanine into dehydroalanine which then conjugates with lysine and cysteine, and in other ways. Treatment with alkali has, therefore, not been seriously studied. Partial acid hydrolysis is an alternative, but this destroys tryptophan and often produces peptides with strong flavours. This is a defect with enzymic digestion also.

Enzymic digests of some proteins, after being concentrated and, if necessary, fractionated, can undergo some resynthesis or rearrangement to yield 'plastein'. The nature of the reaction is obscure, but it can be used to incorporate into the 'plastein' those amino acids that are scarce in the original protein (Yamashita, Arai & Fujimaki, 1976). The potentialities of this procedure are being examined by Savangikar & Joshi (unpublished). It should be possible to make material containing a greater percentage of protein than the original LP because the digest can be filtered to remove insoluble material;

nevertheless, because the final product is insoluble, small molecules that may have unwanted flavours should be removable. An inversion of this digestion procedure is suggested by Staron (1975). Inadequately washed LP was incubated with a fungus (*Geotrichum candidum*) which has little protease activity but contains other active enzymes which would, it was hoped, remove carbohydrates, phenolic compounds and saponins. Staron presents some evidence for the removal of much of the ash and of a protease inhibitor from lucerne LP, and for an improvement in nutritive value. The carbohydrate content was not affected and the amount of LP recovered is not stated. There is as yet no evidence that adequate washing would not have been as beneficial as the fermentation.

One factor leading to the insolubility of many proteins, including coagulated LP, is the approximate equality between the dominant polar groups, —COOH and —NH$_2$. The properties of a protein can therefore be altered by chemically masking one or other type of polar group. Franzen & Kinsella (1976) made a slightly soluble product by treating LP with enough succinic anhydride to cover 84 % of the —NH$_2$ groups. The yields, digestibility and nutritive values of these products have not been reported.

Modified forms of LP will have an enhanced appeal in the conventional food industry because products could be made that disperse easily in water and give textures that are considered desirable. Any such procedure would, however, greatly increase the cost of the product and make production, of necessity, an industrial process. The primary merit of LP, that it can be made economically on a small scale and locally, will be lost if it is extensively modified.

6
Digestibility *in vitro* and nutritive value in animals

If a protein is readily digested by several enzymes *in vitro*, it is reasonable to expect that it will be digested in the gut; if it is not digested *in vitro*, it may still be digested *in vivo* because of the simultaneous action of several proteases and of possible cooperation from the gut flora. The conversion of protein nitrogen to nonprotein nitrogen in silage is well known, the digestion of LP by proteases in several leaf extracts was studied by Singh (1962), and methods for preventing this were discussed in the preceding chapter. It is clear therefore that LP is, in part at any rate, susceptible to proteolysis in its native state. Protein digestibility is usually increased by denaturation: there is therefore good reason to expect digestibility to increase when LP is coagulated by heating. The effects of later treatments may, however, be harmful.

While interest was increasing in the potentialities of LP as a food, there was also increasing interest in the presence of antiproteases in leaves (e.g. Chen & Mitchell, 1973; Richardson, 1977). These, like the more potent antiproteases of legume seeds, are relatively thermostable proteins: one lost only 18 % of its activity after boiling for 6 h (Chen & Mitchell, 1973). Fortunately, extracts from the leaves of spinach and brussels sprouts were much less active than extracts from seeds. Furthermore, antiprotease activity increases in detached tomato leaves within a few hours (Shumway, Rancour & Ryan, 1970), and it is not shown strongly by tomato and potato leaves until they are damaged by insects (Green & Ryan, 1972). There is therefore no reason to expect trouble on that score when freshly harvested crops are used. Nevertheless, compared with casein (which is exceptionally easily digested), even 'cytoplasm' protein is digested slowly *in vitro* by enzymes. The matter merits more attention.

Digestibility *in vitro*

Enzymes of the digestive tract on freeze-dried leaf protein

Akeson & Stahmann (1965) digested 18 samples of LP from 9 species with pepsin followed by trypsin, removed undigested material with picric acid, and measured the individual free amino acids; 10 to 33 % of the amino acids were hydrolysed from 12 conventional food proteins, whereas 19 to 24 % were hydrolysed from the LPs. A similar experiment with larger amounts of enzyme (Bickoff *et al.*, 1975) can be interpreted as showing that 88 % of the nitrogen in LP and 97 % of the nitrogen in 'cytoplasm' protein appeared as free amino acids, but it is not clearly stated that hydrolysis went so nearly to completion. These may be values for soluble nitrogen rather than amino acid nitrogen. From the point of view of nutrition this matters little: peptides as well as amino acids are absorbed.

There is no uncertainty about what was measured in the experiments of Lexander *et al.* (1970) and Carlsson (1975) – it was the percentage of the nitrogen brought into solution by digestion, and also the percentage of that solubilised nitrogen still precipitable by TCA. Using minute amounts of pepsin alone, Lexander *et al.* (1970) found large differences between 23 species. *Amaranthus caudatus* and *Vicia faba* were the most readily solubilised (65 %) and half of the soluble nitrogen could be precipitated by TCA. They found a clear trend: the larger the percentage of nitrogen solubilised, the larger the percentage of it that remained precipitable by TCA. This relationship also held for products of fractionation. 'Chloroplast' fractions were less digestible, and little of what was brought into solution could be precipitated by TCA, whereas digests from the more digestible 'cytoplasm' fractions were 60 % precipitable by TCA. It is possible that this phenomenon could be turned into a useful practical method for making more refined fractions from LP.

In comparisons between 15 species, Lexander *et al.* (1970) found that solubilisation by pepsin followed by trypsin ranged from 84 to 40 % and little material remained precipitable by TCA. LP made from crops given abundant nitrogen fertiliser tends to contain a little more nitrogen than LP from less well manured crops, and there is general agreement (e.g. the papers already cited, and Horigome & Kandatsu, 1964) that the digestibility of LP increases with its nitrogen content. Lexander *et al.* (1970), and Carlsson (1975), working with three species, found a small effect only. Staron (1975) studied the digestion of LP with an extract of rat duodenum (p. 74); the experiments of Shah *et al.* (1967) and Buchanan (1969*a*, *b*) were mainly concerned with the effects of heat treatment and will be discussed later.

Papain on freeze-dried leaf protein

Measurements of digestibility by enzymes from the mammalian digestive tract have an obvious bearing on the potential value of LP as a food for nonruminants. Measurements of digestibility by papain have academic interest and also a bearing on the possible use of modified LP by the food industry. Papain is used extensively and the pawpaw or papaya, from which it is made, is a common fruit in countries where LP could be useful.

Unactivated papain digests LP very slowly; even when activated by potassium cyanide, it digests LP at only one-fifth of the rate at which it digests casein (Byers, 1967*a*). This slow digestion is probably not caused by the presence of inhibitors in the LP because, in its presence, casein is digested at the normal rate. Several publications suggest the presence of antiproteases in LP. These should be regarded sceptically until they are substantiated in experiments with well-washed preparations. Thorough washing is especially important when potassium cyanide is used as the activator because, if reducing sugars have not been thoroughly washed out of the LP, cyanohydrins may be formed, and these yield ammonia during Kjeldahl digestion (Buchanan & Byers, 1969). Cyanohydrins are soluble; they do not therefore cause trouble if the course of digestion is followed by measuring the amount of undigested protein only. It may, however, be prudent to use thioglycollic acid as the activator (Buchanan & Byers, 1969; Saunders *et al.*, 1973; Betschart & Kinsella, 1974*b*) although it is not so effective as potassium cyanide. As in digestion with pepsin and trypsin, part of the LP is made soluble while remaining precipitable by TCA; the amount of material of this type increases as the pH of papain digests is increased (Byers, 1967*a*). The papain digestibility of freeze-dried LP is not increased by extraction with lipid solvents at neutrality and it is diminished by extraction with solvents in the presence of strong acids. This effect is probably caused by hardening and compaction of the particles of LP so that there is less opportunity for access by the enzyme. Buchanan (1969a) restored digestibility by fine grinding and prolonged pre-soaking. Papain, like pepsin and trypsin, digests 'chloroplast' protein more slowly than 'cytoplasm' protein whether the fractions are separated by differential heat coagulation (Byers, 1967b) or centrifuging (Byers, 1971b). Species and age differences in the proportions of the various components of LP may explain some of the small differences in digestibility that Byers found when she compared preparations from 14 species harvested at different ages.

Enzymes on heat-damaged leaf protein

Samples of LP that have been subjected to more prolonged heating than is necessary for coagulation, have invariably been digested more slowly than samples from the same batch not so treated. The extent of the change depends on conditions of heating. Obviously, duration and temperature are important variables. The extent to which air penetrated the mass of LP seems also to be important, as is the period during which the LP was heated in the presence of water. As has already been pointed out (p. 71), a thin layer of moist, crumbled LP, heated in a current of air so that water vapour can escape quickly, probably suffers less than lumps of moist LP in a closed oven even although the oven temperature is lower. Finally, there is the point to which attention cannot be too often directed – inadequately washed LP, heated in the presence of extraneous components of the leaf 'whey', has more opportunity for undergoing detrimental change than more carefully washed material. Although hazardous, heating has compensating advantages; it is a convenient method of sterilisation, and is the cheapest method for making dry products that can be easily dispensed in feeding experiments, or marketed commercially. Much more research is therefore needed on the precise nature of the changes undergone by LP from different sources, and after being pre-treated in different ways, when it is heated. When this work is done, it is to be hoped that the precise conditions of washing, drying and heating will be stated. Several statements about digestibility have not been quoted here because, in the absence of this information, their significance cannot be assessed.

The process of making 'chamois' leather, in which partly dried hide is beaten with the unsaturated fats in fish oil and then heated and exposed to air to oxidise the oil, is so well known that the risk of LP becoming tanned in a similar way was obvious from the start of work on LP extraction. The process probably does not involve the formation of covalent links (Shah *et al.*, 1967; Shah, 1971). There was increasing loss of digestibility when moist LP was dried at 60, 80 and 100 °C. This was interpreted as the result of physical rather than chemical changes in the protein: the loose complex formed between protein and partly oxidised unsaturated fatty acids could be disrupted, with restoration of protein digestibility, by solvent extraction. Oxidation of the fatty acids was inhibited by antioxidants such as amla (*Emblica officinalis*) powder (Shah, 1968).

Three main types of change are relevant when LP is heated during drying: effects of heat on the protein itself, effects on the lipid, and interaction between these two components of the preparation. Buchanan (1969*a*, *b*) dissociated the environmental conditions and measured the extractability of the lipids, the *in vitro* digestibility, and the nutritive value for rats. During storage for a few days at 100 °C or a few weeks at 60 °C with access of air, lipid (as Rouelle had noticed, p. 140) became less readily extractable and the protein less digestible by papain. In conditions in which water vapour could escape freely there was no further change on prolonged heating. Loss of digestibility was obvious in 5 h at 100 °C when LP was heated in sealed tubes so as to retain the water (about 9 %) present in air-dry protein; when only 2.5 % of water was present, several weeks were needed for the same loss. There was a similar loss of digestibility in the absence of oxygen, but solvent-extracted protein did not lose digestibility even in the presence of air and moisture. By solvent extraction after moist heating in a sealed tube, i.e. with restricted access of air, digestibility returned to nearly the normal value even although prolonged heating, especially when air had access to the LP, had made much of the lipid unextractable. The importance of excluding light during prolonged storage has already been commented on (p. 53). The conclusion to be drawn from these observations is that there are advantages in removing water from LP as rapidly as possible although heating in a current of air increases the risk that part of the lipid will be peroxidised. Literature on the toxicity of peroxidised lipids is surveyed by Mead & Alfin-Slater (1966). Speed may be advantageous even if, as Mauron (1970a) suggests with other foods, the changes undergone by lipids affect the digestibility of protein more by mechanical entanglement than by complex formation.

The behaviour of LP when heated, or during prolonged storage, differs in no essential respect from the behaviour of other proteins that contain unsaturated lipids. The behaviour of unfractionated LP when heated after solvent extraction has not been thoroughly investigated: there would be little reason for the investigation because the material would already be dry and sterile. Carefully made preparations of 'cytoplasm' protein are nearly free from lipid. If they were being dried by heating, the possibilities of other reactions, e.g. between the ε-NH_2 of lysine and the β or γ-COOH of aspartic or glutamic acids, and various types of oxidation, would arise. An intermediate preparation, soy LP partly freed from 'chloroplast' protein by centrifuging the original extract at 20 000 times gravity (Betschart & Kinsella, 1974b), was digested more slowly by papain after 24 weeks in air at 27 °C.

Digestibility *in vivo* and nutritive value

Detailed experiments on animals are an essential prelude to any attempt to introduce a novel protein into human diets; they also have direct relevance when economically important animals, such as chickens and pigs, are used. Unequivocal interpretation is, however, extremely difficult. Species differ in their array of gut enzymes, in their amino acid requirements and in their susceptibility to extraneous substances that may be present in the protein. During the phase of work on LP that I have dubbed 'Natural History' (p. 41), every extraction and every feeding experiment was, to some extent, interesting. Now, an extraction is not worth discussing unless the antecedents of the crop are fully described, or a new type of machine was used for the extraction. Similarly, the results of a feeding experiment are now of little interest unless precise information is given, not only about the species from which the LP was made, but, more significantly, about the details of the separation procedure and method of preservation or drying. As the preceding sections show, it is easy to damage a protein by inept handling: poor nutritive value is probably more often the result of technical incompetence on the part of the processor than of synthetic inconsiderateness on the part of the plant. Nevertheless, I have often been amazed at the good results claimed with some products which I knew, having seen the techniques used, to be heavily contaminated or damaged.

Experiments on pigs, chickens and possibly fish (Cowey, Pope, Adron & Blair, 1971) have obvious economic importance. No experiments on chickens have been published since the early results were surveyed by Woodham (1971). Experiments on rats are less relevant. We take little interest in their nutritional welfare, and an animal with a surface to volume ratio differing markedly from our own is a poor model. In a homoiothermic animal that ratio, and the environmental temperature, control the probability that a diet which meets energy needs will also meet needs for specific components such as amino acids and vitamins. However, there is now such a body of knowledge about the nutritional needs of rats that experiments on them deserve attention. The relevance of work on other mammals, and on insects, protozoa and bacteria, is doubtful. Some of them are experimentally convenient, but these organisms are even less likely than rats to have digestive capabilities and amino acid requirements that resemble our own. In this section, results on pigs and rats only will be considered: they do not conflict with results on other organisms.

It is unrealistic to feed an animal on a diet in which all the protein comes from one source. The merits of a protein are therefore usually assessed from experiments in which the protein that is being tested supplies 10 to 20 % of the total protein; the performance of the protein under test then obviously depends on the amino acid composition of the proteins in the remainder of the diet. Performance probably also depends on the precise physiological state of the test animals, on the temperature and illumination at which they are kept, and possibly on the time of year of the test. For these and other reasons, the attempts often made to grade proteins on a numerical scale, sometimes even to three significant figures, seem misguided. It is no more possible to grade proteins in this way than it would be to grade all alloy steels on one numerical scale regardless of the uses to which they are to be put. The impression of imprecision given by vague phrases containing words such as *better, worse* and *nutritive value* will distress orthodox trophologists, but it is less misleading than the spurious precision suggested by a number. A rat is not a chemical reagent.

The first experiment on pigs (Barber, Braude & Mitchell, 1959), which showed that they grew a little faster when a predominantly cereal diet was supplemented at two levels with LP rather than with the same amount of protein in the form of white fish meal, met with some scepticism. Doubts were dispelled by a more elaborate experiment (Duckworth, Hepburn & Woodham, 1961) in which four levels of LP were compared with three of fish meal and one level of groundnut meal mixed with LP. Again, the ratio of live weight gain to food intake was slightly greater with LP than with fish meal. The satisfactory result with the mixture of LP and groundnut was particularly interesting. Lucerne LP replaced soy satisfactorily in the feed of large pigs and, in some experiments, in feed for small pigs also (Cheeke, 1974; Myer, Cheeke & Kennick, 1975; Kanev, Boncheva, Georgieva & Iovchev, 1976; Carr & Pearson, 1976) if carefully made. A commercial product was less satisfactory (Cheeke, Kinzell, de Fremery & Kohler, 1977) and was not eaten as readily as carefully made LP. Experiments in which unfractionated whole juice was fed to pigs will be discussed later (p. 132).

The LP used in some of these experiments was moist press-cake preserved by freezing, or drying in a commercial freeze-drying unit; it is therefore not certain that material of the same quality could be made by less expensive methods of storage or drying. Foot (1974) used the technique of Duckworth & Woodham (1961) and mixed press-cake containing 60 % water with cereals containing 7 to 8 % water; the resulting pellets contained 15 to 16 % water and they stored satisfactorily for several months. By adding propionic acid to mixtures made with cereals of normal moisture content (10 to 12 %), Braude, Jones & Houseman (1977) made pellets that kept well in spite of the presence of 20 % water.

In these experiments there seems to have been no trouble from photosensitisation (p. 48). The early experiments were made with LP from cereals in which there is not much chlorophyllase activity, and the leaf juice was heated quickly (Morrison & Pirie, 1961). But lucerne juice was used in the recent experiments and it was often kept acid for several days before use. It is not immediately obvious why pheophorbide, the photosensitiser, was not formed as in the trials of Carr & Pearson (1974). Pigs in the UK may have been less exposed to light than those in New Zealand: this is a point that deserves fuller investigation.

Almost all that is known about the digestibility of LP *in vivo* and the effects on its nutritive value of different methods of processing and different forms of amino acid supplementation comes from experiments on rats. There is no need to discuss here experiments showing, or purporting to show, differences between LPs made from different species. They may be real: but, as already remarked, until comparisons are made between preparations separated and dried or otherwise preserved, in precisely the same way, from crops taken at several stages of maturity, the reality of species' differences will remain uncertain. The early experiments on rats need not be surveyed again; that was comprehensively done by Woodham (1971). Instead, recent experiments will be considered that have a bearing on the effects of phenolics and other tanning agents on reactions that destroy specific amino acids or make them unavailable, and on complex formation with lipids.

Phenolic compounds and other tanning agents

Phenolics interfere with the extraction of protein from leaf pulps (p. 23). They inactivate enzymes so that elaborate precautions have to be taken if active preparations are to be made from some leaves (p. 57). They tend to depress the nutritive value of fodder for ruminants (McLeod, 1974) but there is evidence that a suitable amount of tannin in fodder prevents the accumulation of foam that causes bloat. Generalisations on these subjects are inadvisable because of the very large number of phenolics that are present in some leaves (Milic, 1972; Thakur, Somaroo & Grant, 1974) and the changes that they undergo during harvesting and processing (Pierpoint, 1971; Van Sumere *et al.*, 1975; Synge, 1975, 1976).

Although people consume many sources of tannin for pleasure, e.g. wine and tea, or of necessity, e.g. sorghum, the main factor besides freedom from fibre, that leads to the selection of certain species as leafy vegetables, is freedom from phenolics and other tannins. By-product leaves of these species will be used as sources of LP but there is no advantage in extracting protein from a leaf that is edible after conventional cooking (p. 42). It is likely therefore that LP will be made from crops containing phenolics because it is easier to get large yields from them than from vegetables – perhaps the phenolics give some protection from attack by pests (Feeny, 1969; Levin, 1971, 1976).

It was reasonable to assume that conjugation with phenolics would diminish the nutritive value of LP, and there is now good evidence that it does. The long series of papers by T. Horigome and his colleagues on the digestibility and nutritive value of casein and other proteins after combination with phenolics need not be surveyed here. These investigations started with work on the digestibility in rabbits of different varieties of clover, and fractions made from them, but most of the publications deal with artificial mixtures, and my knowledge of them depends on abstracts.

A reversible association between phenolics and peptide bonds, which leads to cross-linking between protein molecules, or between parts of one protein molecule, is usually invoked to explain the diminished digestibility of proteins containing phenolics. The evidence is reviewed by Van Sumere *et al.* (1975). It is usually assessed by comparing the amounts of nitrogen in the feces of rats fed on diets containing different proteins, or no protein. The trustworthiness of this method clearly depends on the extent to which different diets abrade the intestinal mucosa, for much of the fecal nitrogen is derived from that source. However, the striking effect of small additions of tannin suggests that the method is valid. Eggum & Christensen (1975) found that 1.5 % of tannin diminished the digestibility of soy bean protein from 93 % to 73 %. After allowing for that diminution, they found that the biological value of the protein was not altered. This became less surprising when they found, by analysing the feces, that it was the inessential amino acids, glutamic acid, glycine and proline, that were rendered least available. 'Chloroplast' protein is less digestible in rats than 'cytoplasm' protein (Henry & Ford, 1965; Subba Rau *et al.*, 1969, 1972); it is reasonable to assume that the greater concentration of phenolics in the former is a partial explanation. It would be interesting to know whether this difference arises because the chloroplasts contain more phenolics, e.g. plastoquinone, or because combination with phenolics makes other proteins coagulate more easily during the heat treatment used for fractionation.

Complex formation with the ε-NH_2 of lysine is the most clearly defined reaction between proteins and phenolics. This, like the comparable reaction between gossypol and the proteins of cotton seed, makes lysine unavailable to nonruminants although it still appears in acid hydrolysates. The amount of lysine bound in this way was measured by Allison (1971) and Allison, Laird & Synge (1973) by destroying with nitrous acid the ε-NH_2s of those lysine molecules that were not complexed with phenolics, hydrolysing, and measuring the lysine that had been shielded from nitrous acid. Allison *et al.* (1973) point out that nitrous acid probably penetrates the protein molecule more completely than the larger molecule of fluorodinitrobenzene, which is often used in a somewhat similar manner, and so will react more completely with unshielded lysine. In lucerne LP which had been protected from oxidative complexing by processing in the presence of sulfite, Allison (1971) found only 6.5% of the lysine shielded from deamination and therefore probably unavailable to nonruminants; 9.5% was shielded in LP made without precautions, and 18.5% when chlorogenic acid was added at the start of the extraction. The values on 15 samples from seven species, made with no attempt to prevent oxidation or complexing, ranged from 13.3 to 38.2%; the larger values were given by 'chloroplast' fractions. The nutritive value of these samples had been measured on rats (Henry & Ford, 1965); it correlated reasonably well with the amount of lysine not shielded from deamination. There is so much lysine in LP (p. 60) that the loss in nutritive value if some of it is made unavailable by complexing with phenolics is not likely to be great. This expectation was borne out experimentally (Henry & Ford, 1965; Shurpalekar, Singh & Sundaravalli, 1969); supplementation with lysine did not improve the nutritive value of rat diets containing LP. Woodham (1971) quotes another trial (Reddy, unpublished) with the same result, but Bickoff *et al.* (1975) found lysine supplementation advantageous.

The measurements made by Subba Rau *et al.* (1972) of the quantity of phenolics in LP preparations have already been referred to (p. 57). They found no correlation between nutritive value and the amount of phenolics in 11 preparations from 8 species, used as sole protein sources for rats. However, when the results were expressed as the ratio of phenolics to nitrogen, it was clear that samples with the greater ratios were the worse nutritionally. Rats did not grow at all when fed on samples with the largest ratios. One of them, made from unwashed carrot (*Daucus carota*) tops, contained 37.2 % ash and only 3.9% nitrogen: there may therefore have been other reasons for the rats' failure to thrive. Omole, Oke & Mfon (1976) found no defect in carrot tops as food for rabbits.

Sequestration of lysine does not seem to be an adequate explanation of the poor nutritive value of some LP preparations. The possibility that phenolics, which are often methylated before being excreted, increase the demand for methylating agents such as methionine, will be discussed later (p. 87). The position will become clearer when there has been more work along the lines started by Davies, Laird & Synge (1975); they hydrogenated a protein: phenolic complex and isolated amino acid derivatives that had been stabilised by the hydrogenation so that they withstood acid hydrolysis. By such means it should be possible to identify the amino acids that have combined with phenolics, and also to identify the particular phenolics, out of the mixture present in a plant such as lucerne, that react most readily with LP. Contrary to the assumption which seems to be made in some publications, all phenolics and other tanning agents are not similarly reactive.

There will be general agreement that complex formation between tanning agents and LP is to be avoided if possible. Tanning agents exist preformed in some leaves. In many more, phenolics oxidise to tanning agents when leaves are pulped in the presence of air. More work is needed to identify the phase of the extraction procedure at which LP is most vulnerable. It would be difficult to exclude air during bulk extraction, but the use of sulfite is conventional in the food industry, is well known to produce pale preparations of LP and would be feasible economically. Such agents must, however, be used judiciously because they can destroy essential amino acids such as cysteine. Undoubtedly, the best policy is to work with species and varieties that are relatively free from phenolics: that would both increase the extractability of LP and avoid damage to what was extracted.

Modification of cyst(e)ine and methionine

A peculiarity of the distribution of amino acids in the bulk proteins from seeds and leaves is that, although minor, individual proteins, such as urease, a protein from the seeds of *Antiaris toxicaria* (Kotake & Knoop, 1911), and the trypsin inhibitors, are among the richest in S-amino acids, a deficiency in sulfur is almost universal. The presence of substances in LP that increase an animal's need for S-amino acids, or the exposure of LP to treatments that diminish the availability of S-amino acids, should therefore be avoided. The detoxication of various aromatic substances by conjugation with cysteine to form mercapturic acids has been known since the end of the last century and may explain the protection given by a high-protein diet against some forms of poisoning, e.g. by *Senecio* (Cheeke & Garman, 1974). The gut flora of herbivores can, to a large extent, metabolise phenolics, but some are detoxicated by methylation. Mice methylate the breakdown products of lucerne tannins (Milic & Stojanovic, 1972). Several authors have made the plausible suggestion that the extra demand for methyl groups, when phenolics are present in the diet, increases the need for methionine. The subject is discussed by Eggum & Christensen (1975); they found no evidence for this effect in rats, though it seems to be real in chickens.

Phenolics may not only increase the need for cyst(e)ine and methionine, they may also make them unavailable. Methionine can react with *o*-quinone (Vithayathil & Murthy, 1972) and *p*-quinone reacts with other thioethers (Bosshard, 1972). There is no evidence for such reactions in the conditions to which LP is exposed during extraction. Quinones and cysteine (Roberts, 1959; Pierpoint, 1969) react in physiological conditions. Cysteine also reacts with plant lactones containing a :CH$_2$ group (Rodriguez, Towers & Mitchell, 1977); this is unlikely to be a problem with LP because these lactones tend to cause dermatitis; farm workers are therefore likely to insist on the elimination of plants containing them from any crop. During preparation, and especially the drying of inadequately washed preparations, LP is probably more susceptible to damage by conjugation with carbohydrates than with phenolics. The formation of complexes partly unavailable to rats was demonstrated with free methionine (Horn, Lichtenstein & Womack, 1968), soy bean (Mauron, 1970*b*) and casein (Pienazek, Rakowska & Kunachowicz, 1975); the reaction has not been demonstrated with LP, but its occurrence is probable.

The amount of methionine in proteins is usually measured after oxidation to the sulfone (MSO$_2$); any sulfoxides (MSO) present are therefore included in the value. Attempts to measure the amount of MSO precisely are frustrated because it is partly reduced to methionine during hydrolysis (Njaa, 1962; Gjøen & Njaa, 1977). Nevertheless, it is worth recording that Byers (in Pirie, 1970*a*) found that 18 % of the methionine appeared as MSO in wheat LP that had been processed quickly, and 30 % in a sample made from juice that had been allowed to stand for 2 h before coagulation. Methionine, and derivatives with a free —NH$_2$ group, are oxidised to MSO by sulfite in air (Yang, 1970); the mechanisms by which methionine in proteins gets oxidised, perhaps enzymically, are unknown. There is also no infor-

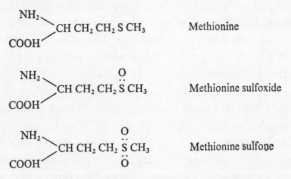

Figure 8. The structure of methionine, methionine sulfoxide and methionine sulfone.

mation about the stereochemistry of the oxidative process. The :SO group in MSO exists in two stereoisomeric forms. In nonenzymic oxidation, both should be formed equally unless the asymmetry at the α carbon atom exerts an influence. One isomer will probably predominate if MSO is being made enzymically. There is no evidence that the :SO isomers differ nutritionally – but the possibility should be borne in mind. They may, for example, have different effects on enzymic hydrolysis of a peptide bond near the methionine.

The nutritional status of oxidised methionine is not completely clear. There is agreement that MSO_2 cannot replace methionine (Miller & Samuel, 1970); much of it is excreted as the *N*-acetyl derivative by rats (Smith, 1972). Conflict over the value of MSO arose mainly from work with chickens. Gjøen & Njaa (1977) found MSO fully equivalent to methionine in rat diets in which there was adequate cyst(e)ine, i.e. the well-known, partial interchangeability of the S-amino acids (e.g. Hove, Lohrey, Urs & Allison, 1974) does not extend to MSO. Byers (1975) argued from measurements of total sulfur that LP from lucerne and lupin contained 2.3 to 2.5 % cyst(e)ine. With these proteins, therefore, the replacement of 30 % of the methionine by MSO should not be detrimental for rats: the value of MSO in human diets is unknown.

In the light of the argument in the preceding paragraph, it seems probable that the formation of MSO is not responsible for the apparent S-amino acid deficiency in LP. Byers (1975) points out that LP from lucerne or lupins contains more than the 3.5 % of S-amino acids that are thought adequate by FAO (1973*a*). Nevertheless, the growth of rats is improved by the addition of methionine to their diets when LP is used as the protein supplement or sole protein source (Henry & Ford, 1965; Shurpalekar *et al.*, 1969; Trigg & Topps, 1971; Hove *et al.*, 1974). A partial explanation is that people, for whom the FAO standard is drawn up, do not need as much methionine in their diets as young rats. This may also be true of pigs; Carr & Pearson (1976) did not improve the rate of growth of pigs by fortifying LP with methionine. There is still no fully satisfactory explanation of the difference in nutritive value between 'chloroplast' and 'cytoplasm' fractions. Subba Rau *et al.* (1972) found less S-amino acid in the former; that seemed to be an adequate explanation,

but Byers (1975) gives reasons for thinking that there is little difference. No one seems to have fed rats on 'chloroplast' protein, at a
level adequate to compensate for its relative indigestibility, supplemented with methionine. The possibility remains that, especially in
'chloroplast' protein, methionine, or cyst(e)ine, undergoes some form
of complexing comparable with that undergone by lysine. The observation that protein made from lucerne that had been soaked in hot
water, or pulped with polyvinylpyrrolidone (which complexes phenolics), contains more cystine than LP made in the normal manner
(Fafunso & Byers, 1977) supports this possibility. But the yields of
protein were not recorded. It is possible that, after these treatments,
only a cystine-rich fraction of the total protein was being isolated.

Unsaturated fatty acids

Although many foods are esteemed because of the flavour of oxidised
fats, it is generally agreed that oxidation is not nutritionally beneficial and may be harmful (Mead & Alfin-Slater, 1966). The conditions in which the lipids in LP oxidise were discussed in the section
on preservation (p. 70); the diminished digestibility associated with
heating and drying has been discussed (p. 79). Buchanan (1969a, b)
found that wheat LP damaged by heat could be restored to nearly its
original state by thorough extraction with lipid solvents. Because fine
grinding had not the same restorative effect, he dismissed the suggestion, made to explain heat damage to fish protein, that the lipids
spread as an impermeable film over the heated particles of protein.
This is not altogether convincing. He found that prolonged soaking
restored digestibility to LP that had been extracted with acidified
solvents: there is no reason to think that sufficiently thorough grinding would not have been efficacious. The point deserves attention
because, although the advantages of working with fresh, moist LP
have been stressed (p. 68), dry material will often be used. During
the short sojourn in a rat's gut, hard dry material may not have time
to soften to the extent necessary for enzyme action. LP may therefore
often appear to have poor nutritive value simply because it was badly
dried. There is no evidence that unsaturated fatty acids combine with
specific amino acids in a manner that makes them unavailable, but
the possibility should be kept in mind.

β carotene and vitamin A

Leafy vegetables are well known as sources of β carotene. Although Hume (1921) had associated greenness with vitamin A activity, and this was confirmed by Dye, Medlock & Christ (1927), it is less widely recognised that only dark green leafy vegetables (DGLV) are rich sources. Illumination seems to be the important factor (Feltwell & Valadon, 1974). Thus the outer leaves of cabbage which are usually discarded may contain 0.58 mg per g wet weight, whereas the 'heart' contains only 0.002 mg (Rothschild, Valadon & Mummery, 1977). Such a dramatic effect of illumination and maturity makes it unwise to trust the values given in food tables for different species: carotene, like other leaf components, is probably more influenced by the physiology than the species of the leaf. There is little or no vitamin A (retinol) in leaves but β carotene is split and oxidised in the body so

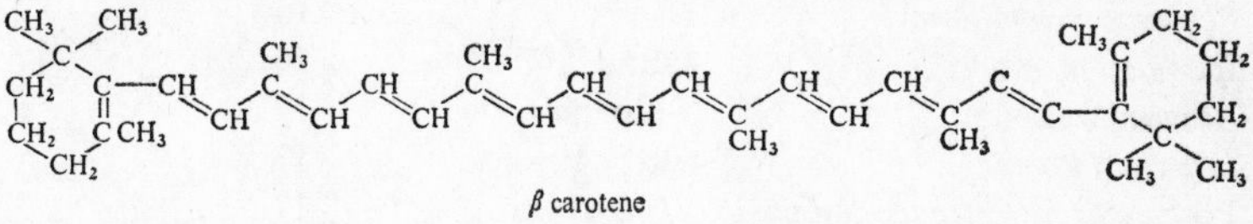

β carotene

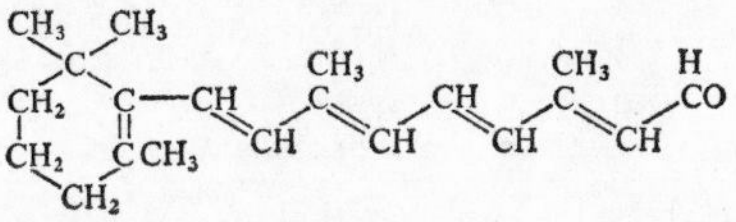

Retinal or retinene

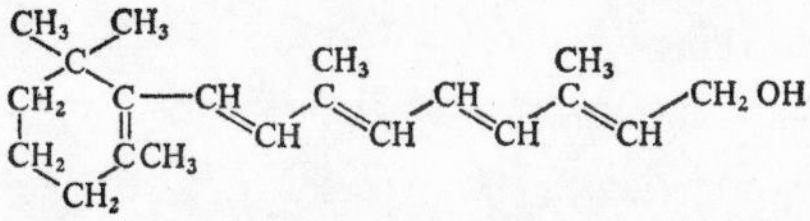

Retinol or vitamin A

Figure 9. The structure of β carotene, retinal or retinene and retinol or vitamin A.

that, in theory, one molecule of carotene yields two of retinol. In practice, scission is not quantitative and absorption is incomplete. Taking these factors into consideration, WHO/U.S.AID (1976) recommends 0.38 mg of β carotene per kg body weight for children and about one-fifth of that amount for adults. Requirement is therefore in the 2 to 6 mg range. A few figures are collected in Table 3 to illustrate the wet weight of leaf needed to supply 1 mg of β carotene;

Table 3. *Some β carotene-rich tropical vegetables. Quoted from A. Pirie* (*1975*) *and derived from FAO* (*1973*b) *and elsewhere*

Latin name	English name	Wet weight of leaf (g) containing 1 mg β carotene
Morinda citrifolia	Indian mulberry	2.7
Carica papaya	Papaya, pawpaw	5.3
Leucaena glauca	Wild tamarind	5.3
Manihot utilissima	Cassava	5.3
Colocasia sp.	Taro	5.3
Sauropus androgynus	—	9
Moringa oleifera	Drumstick	9
Gnetum gnemon	—	9
Hibiscus manihot	—	9
Brassica rugosa	Radish	13
Centella asiatica	—	15
Sesbania grandiflora	—	15
Ipomoea aquatica	Water spinach	16
Amaranthus spp.	'Spinach'	17
Murraya koenigi	Curry leaves	17
Psophocarpus tetragonolobus	Winged bean	17
Vigna sinensis	Stringbean	19
Nothopanax scutellaris	—	19
Momordica charantia	Bitter gourd	20
Ocimum basilicum	Basil	20
Nasturtium officinale	Watercress	20
Basella rubra (*alba*)	Ceylon spinach	24
Talinum triangulare	Purslane	25

Rothschild *et al*. (1977) record that amount in 0.6 g (DM) of nasturtium leaves!

The simplest method for increasing the amount of *β* carotene in the diet, and so preventing 100 000 infants from becoming blind each year because of vitamin A deficiency (WHO/U.S.AID, 1976), is to popularise DGLV. As a first step, more emphasis should be put on the very large yields possible with skilled market gardening (Pirie, 1975a). But there are difficulties. Very young children would have difficulty eating enough leaf to satisfy their *β* carotene requirement unless vegetables from the upper part of Table 3 could be procured throughout the year. In some countries, DGLV are now disparaged – perhaps because they are associated with diets eaten during a more primitive past.

The efficacy of 30 g of DGLV (wet weight and unspecified species) per day in increasing the serum retinol levels of children was demonstrated by Pereira & Begum (1968). The daily consumption of 30 g of leafy vegetable available in South India (Venkataswamy, Krishnamurthy, Chandra & Pirie, 1976) was as effective as massive doses of retinol in clearing the signs of xerophthalmia up to the level of severity called corneal xerosis. Culinary skill is needed to ensure good absorption of β carotene. By measuring fecal carotene and assuming that there is no destruction in the gut (a reasonable assumption that has been made by many since the studies of Hume & Krebs (1949)), Lala & Reddy (1970) deduced that children absorbed 57 to 93 % of the 1.2 mg of β carotene in 40 g of *Amaranthus tricolor* cooked in a little oil. Absorption by adults varied (Rao & Rao, 1970); nevertheless, absorption from DGLV resembled absorption from carrots or papaya. It seems likely therefore that the carotene in LP, which is already finely dispersed, mixed with leaf lipids, and liberated from the cell structure of the leaf, could have an important role in human nutrition. The details of a comparison between LP and some other supplements to the diets of pre-school children are given in the next chapter (p. 109). One of the conclusions may, however, be noted here. The mean serum retinol levels of the groups getting: no supplement, tapioca, horse gram, a cereal+pulse mixture, skim milk and LP were 20.4, 23.3, 34.3, 38.8, 37.6 and 41.3 μg per 100 ml, respectively. It is generally agreed that 20 μg per 100 ml is the value at which signs of deficiency may appear.

As soon as a crop is pulped, the carotenoids are exposed to enzymic oxidation. Enzyme action is particularly associated with the fibre (Arkcoll & Holden, 1973); juice should therefore be separated from fibre as quickly as possible. The pH optimum for the action is on the acid side of neutrality (Walsh & Hauge, 1953; Arkcoll & Holden, 1973); carotenoids are therefore more fully preserved if the crop, or pulp made from it, is made slightly alkaline by exposure to ammonia vapour at an early stage in processing (Spencer *et al.*, 1971). Because of these factors, differences between the carotenoid contents of LP preparations are as likely to be the result of differences in preparative technique, as of differences between the species of leaf. However, preparations made in the same manner in the laboratory from 10 species contained 0.8 to 1.7 mg β carotene per g (Arkcoll, 1973a). LP made in semi-commercial conditions (Miller *et al.*, 1972) from lucerne contained 0.57 mg per g. There is therefore enough β carotene in 1 or 2 g of freshly made LP to meet the daily vitamin A needs of an infant, and enough in 3 or 4 g to meet the needs of an adult.

During commercial production it is most unlikely that there would be 24 hours' delay between extracting leaf juice and coagulating it: in that time, Arkcoll & Holden (1973) found that 19 to 59 % of the carotene was lost at 21 °C. There was little or no loss during coagulation or pressing, but there is a 5 to 15 % loss during most processes of drying. Several semi-commercial methods were compared by Miller *et al.* (1972). Retention after drying depends on the extent to which LP is kept cold and protected from light and air. After a year at 20 °C in air, 90 % was lost; with exclusion of light and air, 12 %; and there was still less loss at −20 °C (Arkcoll, 1973a). The β carotene in LP was as effective as synthetic β carotene for replenishing retinol in the serum and liver of rats that had been on a retinol-deficient diet (Ramana & Singh, 1971). The extent to which loss at various stages in preparation and storage could be minimised by working with stainless steel equipment and nonmetallic containers has not been studied, nor has the protective role of antoxidants such as tocopherol (vitamin E) and the ferulic acid derivative that is present in lucerne (Ben Aziz *et al.*, 1968, 1971). The latter would probably be removed from LP, but not the former. Arkcoll (1973a) found 0.1 mg per g in one sample; Hudson & Warwick (1977) argue that three times that amount of tocopherol would have to be present if it alone were acting as the antoxidant. The behaviour of the other lipid-soluble vitamins does not seem to have been studied.

Supplementation of cereal-based diets with leaf protein

The protein in interesting and palatable diets is, and should be, derived from several sources. There is therefore a lack of realism in experiments in which a concentrate such as LP is used as the sole protein source. When proteins from several sources are being eaten regularly, it is the distribution of amino acids in the whole diet that is important: an essential amino acid deficiency in one component may be compensated by an abundance of that amino acid in another component of the diet. The value of LP as a supplement to barley in pig feeding has already been referred to (p. 82). All the experiments quoted here were on rats.

LP supplemented a rice diet (Sur, 1961), and a simulated Ghanaian diet in which rice was an important component (Miller, 1965). In Miller's experiment it was a marginally better supplement than skimmed milk. That unexpected result may have been caused by the carotene in the LP. An unpublished experiment by Ghosh (in an annual report, and in Pirie, 1970b) deserves mention because LP from water hyacinth, a troublesome weed, supplemented rice when the LP supplied 30 %, 40 % and 50 % of the protein in the diet; its value diminished when more was used. The effect of varying the ratio in which two sources of protein contribute to the total in a diet is clearly shown in experiments by Subba Rau & Singh (1971). In two experiments, the weight gain was greater when half the protein came from wheat grain and half from lucerne LP than with any other ratio. It was argued that this is what would be expected from the amino acid composition of the diets: at that ratio the relative abundance of cystine and methionine in wheat compensates for their scarcity in LP, and the relative abundance of isoleucine, lysine, threonine and valine in LP compensates for the scarcity of these amino acids in wheat grain protein. Supplementation of wheat grain by LP from three species was also observed by Garcha, Kawatra & Wagle (1971) and Kawatra et al. (1974).

Possibly deleterious components in leaf protein

The leaves of many species are well known to be, to varying extents, poisonous. Many species, though they are often eaten in small quantities as vegetables, can cause trouble if they are used as major components of a diet. The possible effects of soluble components in the 'whey' from lucerne will be discussed later (p. 132); attention here is restricted to substances that may remain in LP that has been filtered off and washed.

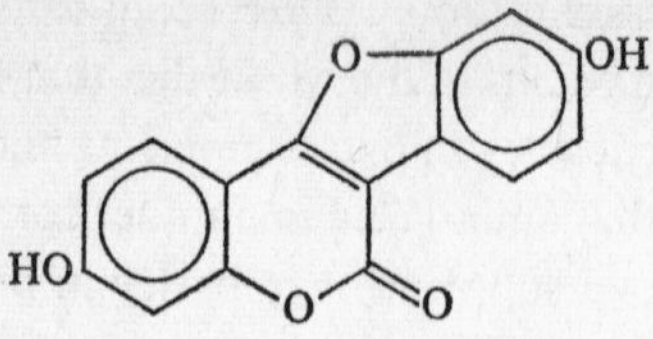

Coumestrol

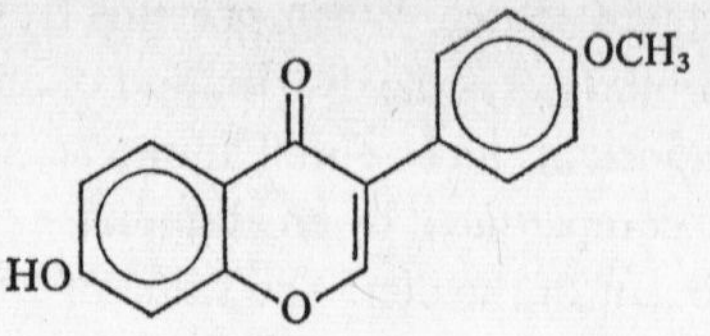

Formononetin

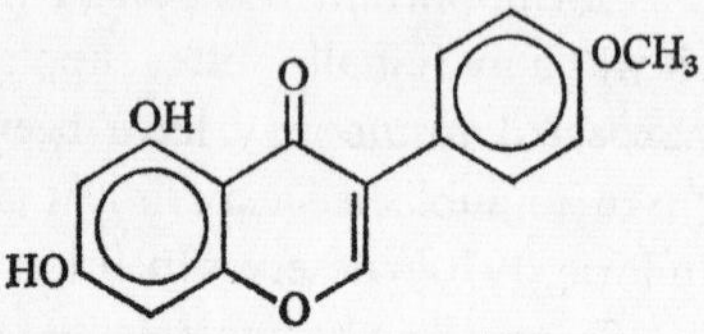

Biochanin A

Figure 10. The structure of coumestrol, formononetin and biochanin A.

LP from a few species, e.g. maize and peas, has a pleasant flavour and can be used unwashed if it is used soon after preparation and before there have been Maillard and similar reactions. Usually, LP intended as human food is washed at about pH 4; that, so far as is known, removes all the alkaloids. Although most of the estrogenic isoflavone in red clover is removed in the 'whey', Glencross *et al.* (1972) identified four isoflavones in washed LP. Two of them, biochanin A and formononetin, were present in quantities that were together equivalent to the presence of 3 μg of diethylstilbestrol in 100 g of LP. If LP from species carrying such isoflavones were being used regularly, methods for destruction, or more complete removal, would have to be devised. Knuckles, de Fremery & Kohler (1976) have already shown that coumestrol, another estrogen, can be removed from lucerne LP by washing at pH 8.5; the final amount present was no greater than in brussels sprouts or peas. Bickoff *et al.* (1975) found a growth depressant in heated lucerne LP that could be removed by an acid wash. There have been several other references (e.g. Ferrando & Spais, 1966) to deleterious components in lucerne.

Several legume seeds contain hemolytic substances that are often classified as saponins; similar substances may occur in leaves. They were found in LP made from quinoa, and *Atriplex caudatus* and *hortensis* (Carlsson, 1975). The quantity increased as the plants matured. There is no reason to think that the amounts present in LP are harmful, but they are among the leaf components that may give LP a bitter flavour.

7
Human trials and experiments

As soon as LP of reasonable quality was being made, the press-cake was regularly eaten as it was being removed from the filter stockings. We had by that time become accustomed to handling green material, and found the product from most species palatable. By 1957 we regularly cooked LP and during the next few years several cooks were employed for short periods so as to get as many new ideas on presentation as possible. For reasons that have already been stated (p. 68), green, moist press-cake was almost always used. The initial disquiet that every cook felt at the texture and appearance of LP disappeared after one or two weeks' experience of handling it.

Within broad limits, the flavour that is acceptable in food is a matter of convention and familiarity. People accustomed to bland foods accept the more strongly flavoured dishes of south India or West Africa slowly; many never accept the somewhat putrefactive flavours of some cheeses and fermented foods. Food with little or no flavour may be considered dull, but it is seldom considered inedible. Fortunately, properly washed LP from many species, e.g. the cereals, cowpea and pea haulm, is nearly tasteless when fresh. A slightly fishy smell, similar to that of China tea, develops as it ages; this is presumably because of the breakdown of choline. Later, oxidation products give dried material stronger flavours. Flavour adheres more strongly to even fresh LP from the clovers, lucerne and potato haulm. It is a pity therefore that lucerne is so often used initially – it imposes an extra barrier to ready acceptance by discriminating adults. However, this may be advantageous when LP is being made primarily as a food for young children. A food that lacks strong appeal for adults is more likely to be reserved for them.

Fresh LP, and LP preserved with salt or sugar, blends smoothly into any mixture; after drying or prolonged freezing it is gritty and has to be ground thoroughly before it is used. Freeze-dried material has a smooth, soft texture, but it is not likely to be used in practice.

Some forms of presentation are obviously unsuitable. Every type of baked product should be avoided because baking enhances the flavours of the breakdown products of the lipids; furthermore, the familiar appearance of bread has so many associations that a greenish version of it is unlikely to be acceptable. LP blends well with savoury flavours, or fish, and less well with coffee, orange or lemon, but banana is an excellent flavouring agent. A mixture of LP and banana pulp has a texture that is well adapted for spreading or insertion into some type of casing.

Little would be gained by devising dishes which contain only 1 or 2 g of LP in a helping: that amount can as conveniently be eaten as green vegetables. On the other hand, LP is intended as a supplement, and it is for many reasons desirable that the protein in a diet should come from many sources. A reasonable helping of the supplemented food should therefore contain 6 to 10 g: a fifth or tenth of the total desirable daily supply of protein. Nevertheless, some of us have eaten 30 g daily for several periods of a week or more; the unabsorbed breakdown products of chlorophyll coloured feces green, but we noticed no digestive disturbance. As already mentioned (p. 10), LP was eaten by a distinguished party in 1940.

The problem of presentation can be considered under three distinct heads. When LP is being given to motivated people, or to people who have accepted the idea that it is a reasonable component of a regular diet, the simplest method of presentation is as good as any other (Morrison & Pirie, 1960). Crumbled moist protein was sprinkled at the table onto a risotto or some similar fairly highly flavoured but protein-deficient dish. The unblended particles of protein are by no means unattractive so long as one has added them oneself; it is more difficult to get acceptance if the mixture is made in the kitchen. In some dishes the protein was made into pieces, encased in thin batter or pastry, that can be eaten without being bitten and examined, or that are bitten only once. For obscure reasons people seem to be less concerned about the internal appearance of small pieces of food than they are about large ones. This can be readily confirmed at cocktail parties or in an Indian or Vietnamese restaurant. It would be interesting, and perhaps useful, to try to make simulated currants that could be put into baked products, and from which the colour would not diffuse out. In that moist form the LP is nearly black. We assumed that the appearance that people accept as normal in a food is purely a matter of convention and that most people would accept the some-what novel appearance of our products after they became familiar with them. Similarly, people habituated to LP accept its flavour so that a larger proportion can be added to a food.

Problems arising under the second head are, to some extent, unreal. Before a novel food can be tried in any country, those responsible for the agricultural and food policy, and representatives of international agencies and other grant-giving bodies, have to be given the food in an attractive form that can be eaten casually during a visit to the laboratory. We therefore devised a set of variously decorated and flavoured morsels, each containing 2 to 3 g of LP. There is no need to describe them here because they have no role in practice, but until such morsels have excited interest, nothing can be done to meet the simpler requirements of the needy. We have not been invariably successful; but only one visitor, unfortunately a senior British civil servant, has made the inane comment 'I prefer beef-steak'.

Work on the third type of presentation was undertaken at a time when most experts considered protein deficiency commoner than energy deficiency. Various dishes were therefore described (Byers, Green & Pirie, 1965) that could be used as protein supplements to accompany diets containing adequate amounts of the other necessary components. The arbitrary standards decided on for these supple-ments were: 25 % of the DM should be protein, and more than 50 % of the protein should be LP. When preliminary trials had shown that a form of presentation was acceptable, a representative sample was freeze-dried, ground and analysed for DM and total nitrogen. The weight of nitrogen contributed by each component in the dish was either determined by analysis, or taken from food tables, or both. These figures enable an estimate to be made both of the final nitrogen content to be expected for the food and also of the proportion of that nitrogen present as LP. We did not accept our estimate of the con-tribution made by LP unless the observed percentage of nitrogen agreed reasonably well with the estimated percentage.

Ten mixtures and methods of cooking were described (Byers *et al.*,

1965) that satisfied these criteria. Two may be given as examples.

Curry cubes. Flour was mixed with a little water to a cream, and boiling water, in which a bag of mixed herbs had been boiling for half an hour, was added slowly to the cream. This was cooked to make a smooth viscous sauce. Curry powder was added to fried chopped onion, and, after further frying, was added to the sauce. Finally sodium monoglutamate and LP were added. After thorough mixing the paste was frozen in trays 1 cm deep and freeze-dried. The cooked flour holds the mass together so that it can be cut into cubes; they keep at room temperature for several weeks and indefinitely in the refrigerator. This mixture is designed for people who like a fairly strong curry. If less flour is used the cubes are more fragile; this does not matter if they are being kept for visitors to sample in the laboratory, but if they are being made for demonstration elsewhere the full quantity of flour should be used.

Contents	Weight taken (g)	Dry weight (g)	Nitrogen content (g)
Flour	28.0	24.40	0.479
Onion	24.0	1.73	0.036
Curry powder	10.0	10.00	0.152
Margarine	20.0	17.25	0.006
Sodium monoglutamate	4.5	4.50	0.336
Barley leaf protein (freeze-dried)	20.0	19.10	2.040
Water	276.0	—	—
	382.5	76.98	3.049

Nitrogen as a percentage of dry matter:
　　Calculated　　= 3.96
　　By analysis　　= 4.23
The percentage of nitrogen due to leaf protein = 67.0

Banana and leaf protein pie. LP was added to mashed banana and mixed well with water. The mixture was put into a pastry-lined tin and covered with a pastry lid; the pie was cooked for 25 min. It can

be served hot or cold. A sweeter tasting mixture can be obtained by
adding sugar to the pie filling.

Contents	Weight taken (g)	Dry weight (g)	Nitrogen content (g)
Banana	150.0	44.0	0.270
Water	150.0	—	—
Short pastry	131.5	95.8	1.262
Mustard leaf protein (freeze-dried)	50.0	48.5	5.580
	481.5	188.3	7.112

Nitrogen as a percentage of dry matter:
    Calculated   = 3.78
    By analysis   = 3.86
The percentage of nitrogen due to leaf protein = 78.5

In Nigeria and other parts of Africa, ground, dried leaves are often
used as a relish; dark green material is therefore not considered
unusual. Methods similar to those described above were used suc-
cessively by Oke (1966, and in Pirie, 1971a). Akinrele (1963), though
he was uncertain about the economics of producing LP in Nigeria,
had no doubts about its acceptability. Dark green vegetables are no
longer esteemed foods in India and Pakistan; when large amounts of
LP are added to familiar foods in these countries, the usual comment
from adults is that the foods taste 'leafy'. In two forms of presentation, LP from cotton was judged better than LP from *Gliricidia*
(Balasundaram & Samuel, 1968). No differences were commented
on when 5 % of LP from three species was added to biscuits (Garcha
et al., 1971). Biscuits are expensive, and baking modifies the lipids in
LP in a manner that intensifies their flavour; this method of presentation is not likely to contribute significantly to the day's protein ration.
Kamalanathan & Devadas (1971) tried several methods of presentation on taste panels in Coimbatore. The two most successful were
'dhal balls' and 'leaves chutney'. The components of the former were
20 g of LP containing 70 % water (i.e. 5 g of actual protein), 22.5 g
of redgram dhal, 1 g each of coriander leaves, curry leaves and green

chillies, 8 g of onion, and 0.5 g of salt. The chutney, or Veppilaikatti, contained 20 g of moist LP, 30 g of bitter lemon leaves, 15 g of green chillies, 25 g of tamarind, 4 g of coriander seed, 5 g of salt, 10 g of oil and 1 g of mustard. Initially the taste panels did not like mixtures containing more than 20 g of LP but, with experience, more was accepted. The experience of Toosy & Shah (1974) was similar.

It seems reasonable to conclude from these trials that when a familiar food has LP added to it and is then judged by a panel, the extent of acceptable divergence from the familiar standard is small. This means that extra care must be taken to ensure that the LP has been adequately washed and pressed, and that it does not have a gritty texture. These points may have been attended to more closely by Byers *et al.* (1965) than by some others who have worked on presentation; that would explain the differing degrees of success achieved. Another important factor is the duration of the trial. Many people react against every novel food on first contact but accept it when it has become familiar – a phenomenon sometimes labelled 'the first day syndrome'. Most of those eating the dishes prepared at Rothamsted were involved in various aspects of the work on LP and had therefore already gained familiarity. This would also be the state of affairs if LP was being produced in a village. I have discussed the interrelations between factors such as these elsewhere (Pirie, 1972).

Work done on presentation in the USA is directed towards producing something that could oust some existing product from the shelves of a supermarket. Such factors as the capacity for holding water or fat, emulsifying power, and the ability to make simulated whipped cream get attention (Wang & Kinsella, 1976*a*, *b*). For these purposes decolourised material is essential, and material separated from a leaf juice by acid precipitation, rather than heat coagulation, is preferable (Miller, de Fremery, Bickoff & Kohler, 1975). What is made by these expensive fractionation techniques is another ingredient that can be added to a conventional food without its presence being noticed; if noticed because of mention on the label, it is assumed that it will be accepted with resignation. This approach does not exploit the basic merits of LP – that it is a human food that can be made by simple processes for local use. Several successes have been commented on, e.g. Anonymous (1974). Betschart (1977) increased to 9 % the protein content of biscuits, that normally contained only 7 %, by adding acid-precipitated protein to them. They were judged unacceptable in the USA when the protein content was increased to 13%. Similarly, she could add 16% of an LP fraction, but not 29%, to a split-pea soup.

In the trials quoted so far, adults were encouraged to express opinions on the merits of the food they were being offered. Many people tend, in these circumstances, to be more critical than they would have been if the food had simply appeared at a meal without comment. Some features that cause adverse comment from a tasting panel might even be regarded as adding piquancy to the dish! Conditions are different in an experiment on adults because, unless they are as strongly motivated as the conscientious objectors used by Hume & Krebs (1949), they must be people confined in a penal or refugee institution. Use in such circumstances gives the food an association with misfortune; a stigma, once acquired, is hard to erase. Furthermore, except in prolonged trials in which epidemiological or work-pattern results are significant, it is hard to find a valid quantitative index of the merits of a food. Adult protein need is sometimes assessed by measuring 'nitrogen balance' – the amount of nitrogen in the food eaten during several days is compared with the amount excreted. It is obvious that the food contains too little protein if the amount of nitrogen excreted exceeds the amount eaten. But there is no reason to assume that the protein intake is optimal as soon as enough is being eaten for a balance to be struck. It is just as reasonable to assume that some excess of circulating amino acids is beneficial.

With children, growth rate is a good index for the merits of a protein supplement; they are the group now most likely to be mal-nourished, they do not have such well-established food prejudices as adults, and those living in institutions get controlled diets which are often meagre or even inadequate. Measurements made on children are therefore valid and relevant. No ethical problem arises: any form of supplementary feeding is likely to be beneficial. The amino acid analyses on LP suggested that it would satisfy human requirements. Table 4 compares the range of values found in many analyses, as well as those already given (p. 60), with suggested specifications (FAO, 1973*a*) for a protein suitable for children.

Table 4. *Amino acid composition of leaf protein compared to the composition suggested by FAO (1973a). (Weight of amino acid (g) in 100 g of protein)*

Amino acid	Leaf protein	FAO suggestion
Isoleucine	4.4 to 5.7	3.7
Leucine	8.4 to 10.7	5.6
Lysine	4.8 to 7.3	7.5
Cystine + methionine	2.7 to 5.2	3.4
Phenylalanine + tyrosine	10.0 to 12.7	3.4
Threonine	4.8 to 5.4	4.4
Tryptophan	2.0 to 3.0	0.46
Valine	5.6 to 6.5	4.1

Quantitative experiments on children (Waterlow, 1962) started 10 years after LP had been eaten regularly by those working with it in Rothamsted. One infant gained 1.2 kg in the 20 days during which half his protein was LP, but the main experimental periods were too short for measurements of growth to be valid; instead, the percentage of the dietary nitrogen that was retained (i.e. that did not appear in urine and feces) when the nitrogen was supplied mainly by milk, was compared with retention when part of the nitrogen was supplied by milk and half to three-quarters by LP made at Rothamsted. Table 5 summarises the results on 10 infants, aged 6 to 20 months, who were part of a much larger group coming into hospital in Jamaica in an acutely malnourished state; their mean weights were only 56 % of the USA standard. It is clear from the table that at the greater intake level (765 and 776 mg nitrogen per kg day), nitrogen was not so well retained from the milk + LP mixture as from milk alone. When the amount of nitrogen in the diet was smaller, though still as great as in most feeding schedules, retention on the two diets was approximately equal. That result agreed with expectation: LP is not as good as milk, but when the supply of milk is inadequate, supplementing the milk with LP is better than relying on a small amount of milk alone. This is shown by the column on the extreme right of table 5 which shows that little nitrogen was retained when the infants were given approximately the same amount of milk as was in the mixture used in the column adjacent to it. There were no problems with acceptance or

Table 5. *Mean values of absorption and retention of nitrogen by malnourished infants at different levels of protein intake from milk (M), or leaf protein and milk mixture (LP + M)*

Diet	M	LP + M	M	LP + M	M
Number of infants	11	5	10	5	5
Weight gain (g per kg day)	5.4	5.2	3.6	3.4	3.2
Energy intake (MJ per kg day)	0.66	0.61	0.53	0.59	0.57
Nitrogen intake (mg per kg day)	776	765	504	496	238
Nitrogen absorbed (mg per kg day)	690	602	452	436	188
Percentage of nitrogen absorbed	89	79	90	88	79
Nitrogen retained (mg per kg day)	276	246	165	160	45
Nitrogen retained as a percentage of intake	36	32	33	32	19
Nitrogen retained as a percentage of nitrogen absorbed	40	41	36	37	33

gastro-intestinal disturbance, but two infants developed an erythema with some swelling of the feet and face. This responded to anti-histamine drugs and was thought to be an allergy. Nothing similar has appeared in later experiments. The most reasonable explanation is that our standards of hygiene were not as high at that early date as they were later.

Growth was the primary criterion used in a more prolonged trial in an institution near Mysore (Doraiswamy, Singh & Daniel, 1969). The normal diet contained ragi (*Eleusine coracana*) flour, beans, vegetables, skim-milk powder, oil and sugar. It supplied 39 g of protein per day and 7 MJ. Eighty boys, six to twelve years old, were divided into four matched groups. One group remained on the institutional diet; one had a daily supplement of 0.5 g of lysine (lysine deficiency is the principal shortcoming of ragi); one group had 24 g of low-fat sesame (*Sesamum indicum*) flour; and one group 15 g of lucerne LP. The last two supplements contributed 10 g of protein to the daily diet and sugar was withdrawn from all supplemented diets to compensate for the energy in the supplement. The results after six months are set out in Table 6. It is reasonable to assume that a group getting milk, if it had been included, would have responded even better than the group getting LP. That control would, however, have been unrealistic in a country where milk is scarce. Nitrogen balance was measured on some of the boys three months after the experiment started. Digestibility on the control and LP diets were 55 and 66 %, respectively, and the nitrogen balances (i.e. amounts of nitrogen retained) were 0.76 and 1.89 g per child per day. The values on the other two diets lay between these values. Taking the groups in the same sequence as in Table 6, and judging by mainly visual criteria, the numbers showing improved nutritional status at the end of the trial were 3, 11, 8 and 13. It is clear therefore that LP is a useful supplement. The boys ate all the diets readily.

Table 6. *Mean measurements on four groups of boys, six to twelve years old, given different diets for six months (from Doraiswamy et al., 1969)*

Dietary supplement	Height (cm)	Weight (kg)	Increase in hemoglobin (g per 100 ml)	Red cell count (millions per mm^3)
None (control)	2.2	0.47	0.29	0.06
0.5 g lysine	4.25	1.05	0.64	0.22
10 g protein in sesame flour	3.51	0.86	0.73	0.19
10 g protein in LP	4.84	1.28	0.87	0.23

In a less fully controlled trial in Nigeria (Olatunbosun, Adadevoh & Oke, 1972; Olatunbosun, 1976) the mothers of 26 children, two to six years old and suffering from kwashiorkor, were supplied with powdered dry LP made from maize and other crops. About 10 g of LP was mixed at home with the usual food of each child. After five weeks, the average total serum proteins increased from 4.9 to 6.6 g per 100 ml. Within 10 days peripheral oedema disappeared, appetite improved, there was less diarrhoea and an increase in mental alertness. Because of the disappearance of oedema, there was at first no increase in body weight, but it increased towards the end of the trial. The authors (in a personal communication) attach particular importance to the prompt recovery from the dejected apathy that is usual in kwashiorkor; the photographs in their paper give striking confirmation of this. Obviously, it is possible that the mothers surreptitiously gave these children other supplements; they were not continuously in hospital. This seems extremely unlikely because, if other supplements had been available, the children would not have got into the original malnourished state.

Because of the success of this trial, pilot-plant production has started using *Celosia argentea*. This is known locally as 'soko'; the LP is therefore called 'sokotein'. It is being given regularly to malnourished children in hospital in Ibadan. *Amaranthus caudatus*, cassava and cowpea are being tried because LP from them has a more attractive colour. The Institute of Church and Society in Ibadan has set up a unit that is ready (July 1977) to produce enough LP to supply 10 g daily to 200 people; it has plans for a unit to produce enough for 1000 people.

In a trial in Lahore (unpublished, personal communication from F. H. Shah), three groups of 10 children were given milk, LP or no supplement. As in other trials, growth, hemoglobin content and appetite improved on the supplements. The results on milk and LP were similar.

The most elaborate feeding trial is not yet completed (June 1977). Only interim reports have been published (Kamalanathan, Karuppiah & Devadas, 1975; Martin, 1975); this account is based on personal communications. Dr R. P. Devadas, Director of the Sri Avinashilingam Home Science College, Coimbatore, became interested in 1966 in the use of LP as a human food, and organised in the college the meeting that resulted in the IBP handbook on LP (Pirie, 1971a). When Mrs C. Martin, the Chairman of the charitable organisation 'Find Your Feet', had collected enough money from various sources to finance a proper trial, Coimbatore, in south India, was the obvious centre for the trial. Because of the very limited amount of control that can be exercised over what people eat, it was clearly impossible to label a properly randomised set of children and feed to each a known diet. Instead, six villages were chosen that are within 9 km of Coimbatore and as similar as possible in character, type of employment and average income. A nursery school (balwadi) was set up in each village, and in this, on six days per week, about 60 children, two to four and a half years old, spend most of the day. On the six school days they get two meals which contain 80 to 90 % of their daily energy and protein supply. For various reasons, mothers were at first reluctant to send children to the balwadis, but once a nucleus of attenders was seen to thrive, recruitment was no problem. Other neighbouring villages are envious of the chosen six. The balwadis have a record of educational and hygienic success that would justify them even if they had no nutritional purpose.

One of the balwadis is primarily educational and the food served in it is modelled, in quantity, character and quality, on food usually eaten at home. In five of the balwadis the food served contains 1.3 MJ more energy than the food served in the control balwadi. In one of these five, most of that energy comes from tapioca which supplies only 1 g of protein. The food given in the other four balwadis contains about 10 g of extra protein given in the form of horse gram, maize + Bengal gram, skim-milk powder or lucerne LP. The energy contents of the supplements are equalised by suitable adjustment of tapioca and sugar (jaggery). So much supplementary energy is being given that the problems tackled by Byers *et al.* (1965) (p. 100) do not arise. The mixture used for the daily helping contains 18 g of lucerne LP, 40 g tapioca, 10 g ragi, 2 g sesame and 30 g jaggery. The moist balls into which it is shaped resemble a sweetmeat (laddu) that is often eaten, and the molasses flavour of the jaggery overpowers the taste of lucerne. There have been no medical problems, and the laddus are so well liked that children who are not among the 60 included in the trial gather round to get any surplus remaining after the 60 have been supplied.

The trial started on 21 July 1975 and will continue until the end of 1977. Inevitably the individuals taking part change; some reach the age at which they attend elementary schools, some leave the region, etc. This complicates clear presentation of the results. The

Table 7. *Interim results of an 18-month comparison of different supplements to the diets of pre-school children*

	Age (months) at the start	No supplement	Diets supplemented with 1.3 MJ		Diets supplemented with 10 g protein		
			Tapioca	Horse gram	Maize + Bengal gram	Milk	Leaf protein
Increased height (cm)	24–30	9.6	10.0	10.3	10.8	12.0	10.6
	36–42	9.3	9.4	9.8	10.3	11.8	10.3
	48–54	8.8	8.3	8.9	10.1	10.8	10.1
Increased weight (kg)	24–30	2.7	2.8	3.0	3.0	3.4	3.2
	36–42	3.0	3.1	2.9	3.3	3.6	3.2
	48–54	2.9	2.9	3.0	3.0	3.4	3.0
Increased hemoglobin (g per 100 ml)	24–30	2.3	2.1	2.3	1.6	2.2	2.5
	36–42	2.4	2.9	2.4	3.1	3.2	3.5
	48–54	2.4	2.5	3.0	3.7	4.3	4.2

children are weighed and measured every three months and from time to time other measurements, e.g. retinol (p. 93) and hemoglobin, are made on those willing to be pricked for a blood sample. The general clinical picture is also assessed from time to time; there was improvement in all the balwadis; it was particularly striking in the one getting LP. The final results will not be assessed until 1978: Table 7 is a summary of the results after 18 months. As was to be expected, milk is a better supplement than LP, but LP is marginally better than the other two sources of protein. Little emphasis should be put on that point. The object of the trial was not to arrange the locally available protein sources in order of merit, but to see whether LP should be classified among the acceptable and useful local sources. It is clearly established that it should be so classified. Apart from establishing the value of LP, the main points that arise from the trial are that children readily accept LP laddus, and their mothers are unperturbed by the greenish tinge of their feces. Another point worth noting is that the group getting extra energy without extra protein shows little improvement over the control: this tends to contradict the statement that is now often made that lack of energy is more common in regions such as south India, than lack of protein.

The preliminary results in Coimbatore are so encouraging that a trial on 600 children is being planned. With support from a charitable organisation in the Netherlands, a similar but smaller trial is proceeding near Aurangabad (unpublished information from Dr R. N. Joshi). Children will probably be the recipients of most of the protein supplements produced from various sources. Trials such as those in India, Nigeria and Pakistan will therefore be followed with the greatest interest.

Note added in proof. The Coimbatore trial is now completed. The picture that emerges from the final analysis of the results is similar to that emerging from the preliminary analysis. The most significant difference is that tapioca is a little more beneficial than it appeared to be at first.

8
The value of the extracted fibre and the 'whey'

When a raw material is fractionated, there is sometimes little uncertainty about which fraction is the primary product and which the by-product. Sometimes, the by-product is treated as a waste either because of technical difficulties, habit or laziness. More often, all the products of a fractionation are recognised as having value, but the value put on each may be arbitrary, or may depend on the demand for that particular type of product at the time. Though not universally understood, these points have often been made before. Thus Lawes (1864) said: 'Linseed and other cakes are themselves, in one sense, manufactured foods. But the object of the manufacturer is not the production of cake, but of oil. If the farmer did not use the cake at all, it would still be made, and the oil would be sold for a higher price. As it is, the manufacturer makes the cake as a by-product, and the price he gets for it enables him to sell his oil so much the cheaper. But if manufactories were set up for the special purpose of preparing foods for stock, the whole cost of the undertakings must be charged upon the food . . .' This principle applies with particular force to fodder fractionation. The DM of an average crop is distributed between the three products – LP, fibre and 'whey' – in approximately the ratios 1 to 5 to 1. Different crops and methods of fractionation give different ratios, but it is obvious that uses must be found for all three products, and it is possible that each will be more valuable when separated from the others than in the original mixture. The value of LP as a food for nonruminants is the main theme of the preceding chapters. The value of the fibre may be enhanced, although it contains less protein than the original crop, because it is drier and has an altered texture. The components of the 'whey' may be more valuable as substrates for microbial growth than as animal feed. After toxic or unpalatable material has been removed from such protein sources as potato haulm, the residual fibre becomes a new source of ruminant feed. Furthermore, unless all the products of a fractionation are used, their disposal would raise acute problems.

The patents taken out by Ereky (1927) and Goodall (1936) refer vaguely to the conservation of the fibre from which they had extracted juice. Better silage can be made from lush crops if they are wilted and bruised so that there is less loss from seepage (e.g. Derbyshire, Gordon, Holdren & Menear, 1969), and quicker compaction of the mass: it is obviously more economical to conserve a crop by drying if pretreatment has removed some of the water. During the 1930s several attempts were made to 'dewater' crops with various commercially available machines – including the now-fashionable twin-screw expeller. They were all unsuccessful because of inadequate appreciation, shown by the repeated use of such phrases as 'pressing out juice', of the fact that pure pressure gets little juice from an un-rubbed leaf. Because the machines used did not apply pure pressure, there was some rubbing; juice was therefore liberated, and protein was liberated along with it. This protein was recovered by Dawson (p. 14) and Goodall (1950) but was wasted by Randolph, Rivera-Brenes, Winfree & Green (1958) and Casselman, Green, Allen & Thomas (1965). If oil prices had increased at that time, and if there had then been the present sensible concern with waste of energy, the primary product of fodder fractionation would have been the fibre: LP would now be treated as the by-product. In countries such as Britain, this may already be the position. Lack of appreciation of the value of the fibre explains the various gloomy guesses at the cost of LP production that were made 10 or 20 years ago.

Fodder fractionation may enhance the value of an existing crop; it also gives farmers a motive for increased productivity. When forage will be used as feed for ruminants, there is little incentive to fertilise, harvest and irrigate so as to get maximum yield. The material harvested would contain more protein than a ruminant needs and would be so lush that it would be difficult to conserve. The argument that a national policy of LP production would increase the amount of cattle fodder, because the incentive to produce protein-rich leaf would of necessity increase the amount of fibre DM produced, was thought paradoxical when first advanced (Pirie, 1942*a*), but it seems to be valid. In Britain 39 % of the permanent grass gets no fertiliser

(National Economic Development Office, 1974); most of this land is probably too steep or rough to be mown, but much of the grass on land that is, or could be, ploughed would yield more DM if more adequately fertilised. A farmer can make full use of a crop that contains more protein than a ruminant needs, by mixing it with a digestible but protein-deficient material such as beet pulp or alkali-treated straw. But the lushness of a well-fertilised crop would remain a problem if fodder conserved by drying were required. This problem becomes particularly acute when a drier is operated continuously so as to make maximum use of equipment. A crop may then have to be processed although it has dew, or even rain, on it. In these conditions, mechanical removal of water is essential for economy.

Most of the crops that are dried commercially contain 25 to 14 % DM; with surface water they may contain only 7 %. When the DM is as small as 10 %, direct drying becomes uneconomic and the crop is allowed to wilt in the field before being dried. There may be little loss if, by chemical treatment, the standing crop is wilted before being mown. If it is mown, left on the ground to wilt, and then collected, there is considerable loss because of incomplete collection (e.g. Gordon, Holdren & Derbyshire, 1969). To put these figures into perspective: to get 1 t of DM from crops initially containing 7, 10, 14 and 25 % DM it is necessary to remove 13, 9, 6 and 3 t of water, respectively. Fibre coming from an efficient LP production unit contains 30 to 40 % DM. To get 1 t of DM from such materials, only 2.3 and 1.5 t of water must be removed. During periods of fine weather it is possible to remove that amount of water by drying in unheated air. The final product of any form of commercial drying contains 10 to 14 % of water; if calculations are made on that basis, the apparent advantage of putting pressed fibre into the drier becomes even greater.

Early interest in fodder fractionation was aroused by the possibility that half to three-quarters of the evaporation load could be removed from the drier, and this possibility was part of the justification for repeated reemphasis of the potentialities of such processes (e.g. Pirie, 1953, 1966b). It may not be possible, with existing equipment, to produce fibre that consistently contains as much as 40 % DM, and it may not be advisable to do so because of the approximate proportionality between the removal of water and the extraction of protein. Although protein is more valuable when extracted as food for people and other nonruminants, than when left in the fibre as ruminant feed, there seems to be a consensus that the protein content of the fibre should not be diminished too much. Unless there is surface moisture on the crop, the difference between the water content of the crop and the fibre has usually been 10 to 15 % in large-scale work. The economic advantage of this type of fractionation process, depends on the amount of energy needed in a mechanical process compared with the amount needed for evaporation, and on the value of the fractions made.

The quality and use of the fibre

All the properties of the fibre depend on the quality of the crop that was processed and the thoroughness with which it was extracted. Byers & Sturrock (1965) found 0.74 to 3.3 % nitrogen in the fibres from 17 crops harvested at different ages and fertilised at different levels. Fibre from hybrid napier grass (gajraj) contained 1.6 to 2.2 % nitrogen (Gore *et al.*, 1974) and from berseem 1.9 to 2.8 % (Mungikar *et al.*, 1976a). The larger values were given by crops that initially contained the most nitrogen, but there was no strict proportionality. During the extraction of LP, salts, carbohydrates and other substances are extracted. Consequently, if half the protein is extracted from a crop, the nitrogen content of the fibre is not halved. Table 8 shows the extent to which nitrogen was removed from several crops by extraction.

For three reasons the value of the fibre as a feed for ruminants is likely to be greater than the value of a crop that initially contained the same amount of nitrogen as these samples of fibre. The fibre is made from younger and less lignified leaves, a larger fraction of the nitrogen in the fibre is true protein, and the comminution of the fibre increases its palatability. Furthermore, ruminants make less efficient use of soluble proteins, which are hydrolysed and partly deaminated in the rumen, than of protein that, for various reasons, is insoluble and by-passes the rumen (Chalmers & Synge, 1954; Kempton, Nolan & Leng, 1977). This gives fibre-bound protein an added merit. In the

Table 8. *Nitrogen contents of the original crop and of extracted fibre made from it*

Crop	Original percentage of nitrogen in the dry matter	Percentage of nitrogen in the fibre	Reference
Lucerne	3.1 to 3.5	2.5	Hartman *et al.* (1967)
Pea vine	2.0	1.55	Hartman *et al.* (1967)
Average of five grasses	2.9	2.2	Maguire & Brookes (1973)
Lucerne	3.5	2.7	de Fremery *et al.* (1974)
Lucerne	4.8 to 2.5	4.3 to 2.4	Connell (1975)
Ryegrass (average of five harvests, May to October)	2.6	2.0	Connell & Houseman (1977)
Lucerne (early)	3.3	2.4	Connell & Houseman (1977)
Lucerne (late)	3.2	2.5	Connell & Houseman (1977)

1950s we found that cattle ate fresh fibre readily. This point has been stressed by Greenhalgh & Reid (1975), Houseman & Connell (1976) and Connell & Houseman (1977). In a comparison between fresh grass and the extracted fibre made from it, the daily live weight gain of cattle was 0.73 and 0.84 kg, and the weight (DM) of fodder eaten was 6.78 and 6.26 kg. That is to say, the weight of grass needed for 1 kg of live weight gain was 9.28 kg (DM), whereas the weight of fibre needed was only 7.45. Other similar comparisons have been published but in less detailed form. The digestibility of the fibres, whether measured *in vitro* or *in vivo*, obviously depends on the quality of the original crop; it also depends on whether the measurements are made on fresh, moist fibre or on fibre that has been dried. Because of the preferential removal of the more digestible 'cytoplasm' protein, there is always some loss of digestibility – about 5 % with fresh material and 10 % with dried (Vartha, Fletcher & Allison, 1973; de Fremery *et al.*, 1974; Greenhalgh & Reid, 1975; Houseman & Connell, 1976).

Pressed fibre has been dried satisfactorily in various types of commercial grass drier. Dry material is a convenient component of cattle feed, but even after part of the juice has been removed, drying is still expensive and is often an extravagance. As soon as there was regular LP production at Rothamsted we made silage, with the fibre from several species, in concrete drain pipes (60 cm diameter) and in smaller vacuum packs. Enough 'whey' was added to the fibre to saturate it, but not enough to produce any effluent. The precise quan- tity added did not seem to be critical and cattle ate the silage readily. It is not easy to understand the failures that have been reported (e.g. Raymond & Harris, 1957). Vartha *et al.* (1973) made unsatisfactory silage until they added molasses and acid; they suggested that the silage contained less calcium than was desirable for sheep feeding. This also is not easy to understand; Braude *et al.* (1977) suggested that the trouble experienced in some trials of whole leaf juice as a pig feed was caused by excess calcium. Fuller investigation of the effects of different degrees of extraction, and of reintroducing some of the cations and carbohydrates by adding some 'whey', is clearly needed. No difficulties were reported in making silage from the leaf fibre of beet, carrot, beans, potato and lucerne (Oelshlegel *et al.*, 1969); silage from the fibre of mixed water weeds after extracting LP from them was rejected by cattle. Silage has also been made from berseem (Magoon, 1972), hybrid napier grass (Mungikar & Joshi, 1976) and lucerne (de Fremery *et al.*, 1974). Some of these authors commented that the fibre made better silage than the original crop. Wallace (1975) quotes the suggestion that there may be so much protein in unprocessed mixtures of ryegrass and clover grown in New Zealand that the silage is buffered at too great a pH for proper fermentation; partial removal of protein would then be advantageous.

The possibility has been mentioned (p. 56) that some types of leaf that are unpalatable or toxic could be used as sources of LP. Processing removes much of the soluble material; the fibre may then be palatable and safe. If there is any doubt about its safety, e.g. with potato haulm, the fibre should be slightly acidified and then re-extracted. This is a matter for experiment with each species.

Cattle are accustomed to diets containing much carotene, and there is evidence that carotene deficiency leads to infertility. Although most of the carotene associated with LP is stable when the LP is properly processed (p. 94), the carotene in the fibre is rapidly destroyed (Arkcoll & Holden, 1973). Whether the fibre is being preserved by drying at a high temperature or by making silage, if the product will be an important component of the total fodder given to cattle, it may be advisable to start conservation quickly so that enzymes are inactivated by heat or prevented from acting by the absence of oxygen in packed silage.

The quality and use of the 'whey'

The deproteinised juice from an LP production factory, which by analogy with cheese manufacture is often loosely called 'whey', would have the Biological Oxygen Demand of the sewage from a small town. It would therefore be extravagant to discharge it into a sewer or treat it so that it can be discharged into a stream. Of the three products of fodder fractionation, this is the most variable; the DMs of 61 samples from nine species ranged from 11 to 47 g per litre (Festenstein, 1972). If crops harvested in wet weather had been included, there would have been 'wheys' containing less DM. When LP is being made as a human food, the water used to wash it would be added to the 'whey' and would nearly double its volume. A comparable amount of water would be used for cleaning the equipment, and that might be added also. The values suggested in the following paragraphs are based on analyses of true 'whey' and do not make allowance for this probable dilution.

The concentrations of the various substances dissolved in 'whey' span an even larger range than the DMs. Thus the nitrogen contents were 0.25 to 1.2 g per litre, and the total carbohydrate contents were 2.5 to 22 g per litre. The true protein content of correctly heated and filtered 'whey' is negligible. Festenstein (1972) measured the contribution made by glucose, fructose, xylose, sucrose, fructose, oligosaccharides and fructosan to the total carbohydrate; there were great seasonal and species differences. Thus glucose was a negligible component of the total in 'whey' from ryegrass harvested in August, but made up 28 % of it with wheat in June; at that time, 13 % of the carbohydrate in 'whey' from wheat was sucrose, but it was never found in mustard. Allowance for this variability must be made when suggesting uses for the 'whey'.

The simplest way to use 'whey' is to carry it back, as a return load, to the fields from which the crop was harvested. About half the wet weight of the crop is in the 'whey'; it will therefore contain nearly half the potassium and a smaller fraction of the phosphorus in the crop. 'Whey' from a leguminous crop may contain a quarter of the crop nitrogen; the fraction is smaller with cereals unless they are harvested soon after being fertilised. No systematic analyses for these elements have been published; their actual value as fertilisers can only be guessed. Using 1977 prices, a reasonable guess at the value of a ton of 'whey' as fertiliser is 10 to 20 £. If 'whey' were restored evenly to the full area from which the crop came, it would form a layer only 1 to 3 mm deep and would supply a substantial part of the potassium needed for the growth of a second harvest.

As is well known, germination and plant growth may be inhibited on land heavily contaminated with silage effluent. Some plant extracts have a general phytotoxic action; with others, phytotoxicity is more specific (e.g. Marchaim, Birk, Dovrat & Berman, 1972; Allison, 1973; Moore, 1976). There was therefore some fear that 'whey' would be harmful. If returned to an area much smaller than the area from which the crop came (the conditions in the neighbourhood of a silo), it inhibits germination and may 'scorch' young leaves. So far as we know (Arkcoll, 1973*b*) it is harmless when returned to as little as a fifth to a tenth of the area from which it came. This was also the experience of Ream, Smith & Walgenbach (1977). By encouraging the growth of gas-forming bacteria such as *Clostridium butyricum* and *Clostridium pasteurianum*, the carbohydrates in 'whey' can improve the structure of intractable soils; this effect persists after the gases escape because of soil stabilisation by bacterial polysaccharides (Arkcoll, 1973*b*).

Most of the components of 'whey' are of doubtful nutritional value for nonruminants, and 'whey' is so dilute that it would be difficult to dispose of all of it in ruminant feed in its original state. Hollo & Koch (1971) concentrated it *in vacuo* and mixed the resulting paste with the

LP. The advantage of complicating the process in this way, rather than spray-drying the whole juice as Hartman *et al.* (1967) suggested, is that the 'whey' does not coagulate on heating and so can be concentrated in a thermally efficient, multiple-effect or vapour-compression unit. In spite of the economy thereby attainable, it seems unlikely that it would be worthwhile evaporating 'wheys' at the more dilute end of the concentration range. If 'whey' were to be concentrated, it should not be added to the LP, but should be incorporated in the fibre and used as ruminant feed. When 10 % of a cattle diet consisted of concentrated 'whey' rather than molasses, growth rate increased (Bris *et al.*, 1970).

The rapidity with which 'whey' stinks when left for one or two days in summer shows that it is a good microbial culture medium; if used in that way it would not need to be concentrated. This use was suggested as soon as systematic work started (e.g. Pirie, 1951) and samples were sent to various organisations interested in the cultivation of microorganisms, but its potentialities have not yet been thoroughly investigated. That cannot be expected until a regular supply is assured. Jönsson (1962) compared as substrates for seven microorganisms, 'whey' from pea vines with four other agricultural byproducts. *Rhizobium meliloti* grew particularly well on it; when an extra source of nitrogen was added, penicillin production was satisfactory.

Later studies have been concerned with the production of 'biomass' that could be used as human or animal food. The 'whey' from 1 kg of water hyacinth yielded 12 g (DM) of a strain of *Saccharomyces cerevisiae* isolated from rotting hyacinth (Oyakawa, Orlandi & Valente, 1968). This may show the advantage of using a yeast strain already adapted to the substrate. Butt, Ahmad & Shah (1972) got only about half that yield from nine strains grown on 'whey' from berseem, mung bean (*Phaseolus mungo*) or guar (*Cyamopsis psoralioides*). Paredes-Lopez & Camagro (1973) found four yeasts that grew well on lucerne 'whey' and three that did not. The sample of lucerne 'whey' used in experiments in New Zealand (Barnes, 1976) still contained 3 to 5 g of protein per litre. Both *Saccharomyces* sp. and *Rhodotorula* sp. converted half of that, and of the 0.9 to 1.5 g of free amino acids per litre, into 5 to 6 g of 'biomass'. The carbohydrate in the 'whey' was used up completely. Results with *Aspergillus, Pseudomonas, Flavobacterium* and *Corynebacterium* were similar. Worgan & Wilkins (1977) quote an experiment showing that *Fusarium semitectum, Trichoderma viride* and *Aspergillus oryzae* removed 65 to 78 % of the Chemical Oxygen Demand from lucerne 'whey'. A succession of organisms will probably make even fuller use of the components of 'whey'. Thus Kummerlin (unpublished) got 12 g of 'biomass' from *Candida utilis* followed by *Endomycopsis fibuliger* growing on a litre of 'whey' from tall fescue (*Festuca arundinacea*); the yield was smaller when the sequence was reserved. In Hungary, a litre of 'whey' (species unspecified) yielded 25 to 30 g (DM) of yeast (Koch, 1974).

These results serve only to illustrate a potentiality. It would be reasonable to use an organism, such as *Candida*, that can hydrolyse polysaccharides. For practical convenience the funguses that form mycelia have advantages because the 'biomass' can be collected by filtering rather than by centrifuging. *Paecilomyces* and *Rhizopus* seem not to have been tried. The latter is particularly worthy of trial because it is already eaten extensively in South East Asia. Furthermore, fungus mycelia tend to contain less nucleic acid than yeasts. The differences in composition that must be expected between different batches of 'whey' undoubtedly complicate their use as microbial substrates. With experience, the differences between successive harvests can be predicted from results such as those of Festenstein (1972) and of Nowakowski & Byers (1972), who studied the effect of different fertiliser regimes on the nonprotein nitrogen of ryegrass. The necessary additions, usually nitrogen and phosphorus, needed to produce a balanced medium can then be made by rule of thumb. Even if the primary object of microbial cultivation is to diminish the Biological Oxygen Demand, some chemical additions will probably be needed because 'whey' usually contains excess carbohydrate. The use of 'whey' as a microbial substrate is the ultimate ideal, but the best short-term measure is to put 'whey' back on the land. Irrigation is needed in eastern Britain in four summers out of five.

Plant extracts comparable to 'whey' are well known as sources of substances with proved pharmacological activity. Nevertheless, claims that 'whey' stimulates plant growth to an extent that cannot easily be explained as the result of the presence of nitrogen, phosphorus and potassium, etc., or that 'whey' contains 'unidentified growth factors' for animals, may be treated with some scepticism. The roster of physiologically active molecules is probably far from complete, but mortality among claimants for inclusion in that roster is large. For example: 11 authentic B vitamins no longer precede vitamin B 12. It is, however, worth recording that the triacontanol in lucerne extracts stimulates the growth of several plants (Ries, Wert, Sweely & Leavitt, 1977). McClure (1970) gives an enormous list of substances, many of them pharmacologically active, in water weeds; and water hyacinth 'whey' may stimulate plant growth.

When the 'whey' from a certain species contains a substance of proved efficacy, as a growth stimulant, it could be used as a spray. The exploitation of 'whey' as a source of purified substances depends on those substances having properties that are sufficiently unlike those of unwanted constituents of the juice for separation to be possible by commercially viable techniques. Attempts to separate ascorbic acid from lucerne juice failed several years ago because the ratio of ascorbic acid to the total quantity of organic acid collected on ion-exchange resins was unfavourable. However, if potato haulm were being used as a source of LP, it would be simple to precipitate the glycoalkaloids (e.g. solanin and related substances), and little else, from the 'whey' by adding alkali. There is at present little demand for these substances – that was at one time the situation with the steroids in agaves and yams which are now important as starting materials for pharmacological syntheses. Alkaloids can be separated either by precipitation or by extraction into a solvent immiscible with water. Skinner (1955), Mitscher (1975) and Roja & Smith (1977) list many plants with antibiotic activity – though the root rather than the leaf was usually studied. It is even possible that dyes, such as indigo and madder, could be by-products of LP production and could emulate rubber by competing with synthetic material.

These suggestions may seem like 'turning the clock back'. It is worth bearing in mind that industrial chemistry prevailed over extraction from 'natural' sources at a time when a science-based industry contended with traditional, or even peasant, agriculture. The balance may sometimes tilt the other way when agriculture also has a scientific basis. Furthermore, there are now proposals that photosynthetically generated 'biomass' should be used as a source of energy for heavy industry. If that proposal were followed up, some of the energy would be used to drive chemical synthesis. It might be simpler and more logical to let the plants do the synthesising.

9
The role of fodder fractionation in practice

Evidence set out in the preceding chapters shows that the production of LP in bulk is technically feasible and that the product is nutritionally valuable. Stress was laid on the importance of considering the whole fractionation process without assigning preeminence to any one fraction. It is worth recording that in about 1960 a research group in Florida was collecting the fibre and discarding the juice; a group in Uppsala was collecting the 'whey' and discarding the fibre and LP; in Rothamsted we collected the LP and discarded the fibre and 'whey' but emphasised in all publications that this was a temporary state of affairs. There is nothing novel in the idea of fractionating plant material. Cereals are separated from chaff and often from bran. Potatoes are peeled. Sugar and oil are extracted from various crops. The petioles and outside leaves of many vegetables are rejected. It is odd, in the light of the last example of separation or selection, that it has taken so long for the idea of fractionating forages to gain acceptance. Our Neolithic ancestors lacked the technical ability to do this, and therefore subjected forages to conversion in animals rather than to separation; we may have relied on their methods too exclusively and for too long.

The reasons for thinking that more attention should be paid to crops yielding leaves rather than seeds or tubers, and that leaf crops should be fractionated, may be restated:

1. Between 10 and 30 % of the protein in a forage is converted into human food by ruminants, whereas 40 to 60 % of the protein can be extracted.
2. Leaves are the main site of protein synthesis and there are losses during translocation to other parts of a plant.
3. When LP is made, the crop is harvested when less mature than when silage is made, and when much less mature than when hay is made or a conventional crop is taken; the cost of harvesting is

greater but an immature crop is not at risk for so long from diseases and pests.

4. Crops that regrow several times after being cut young, or perennial crops, maintain cover on the ground; this enables fuller use to be made of sunlight and protects the ground from erosion.
5. The fibrous residue contains the protein that was not extracted. Depending on the processing conditions, it can have two to five times as great a percentage of DM as the original crop and can therefore be dried to produce conserved ruminant feed more economically.

Clearly, there are many reasons for advocating fodder fractionation and many situations in which it could be embarked on in practice. This, unexpectedly, arouses scepticism. An advocate is often asked whether factory, farm, or family operation is envisaged. The reply that all three procedures are suitable and that choice depends entirely on the particular situation seems to some sceptics as unconvincing as the advertisement for a panacea. Similarly, there is often argument about the relative importance of LP as a human or an animal food, and about the extent to which it is advantageous to deplete the fibrous residue of protein. Again, the answer depends on circumstances. In a poor country with poor communications, where it rains nearly every day, and where protein sources are scarce, it would be advisable to extract as much LP as possible and eat it near where it is being made. In an affluent country with a long winter during which cattle need fodder, attention would be paid to the partial dewatering of the fibre rather than to the extraction of LP, and most of what was extracted would be fed to pigs or poultry.

There will be little dispute about the value of a policy of fodder fractionation when the starting material is at present abundant and unused, e.g. a water weed or a by-product. There may be doubts about the feasibility of collecting material from water, or from the scattered sites at which by-products accumulate, but there will be no doubts about the advantage of making fodder and LP from an unexploited resource if that is feasible. But sites where a supply of weed can be regularly collected are, fortunately, rare, and by-products are intermittent. The main questions that arise concerning fractionation techniques are: 'Should a crop that is already being used unfractionated, be fractionated; and should a crop be grown for fractionation in place of a crop grown for conventional use?' These questions can be considered either in terms of the expected yield of edible proteins, or in terms of profitability.

Local production and use

The simplest situations to consider are communities in which most of the food is produced locally for local consumption. If, because of a local resource such as fish, protein is not scarce, there would be little advantage in making LP. There, if food is scarce, energy will be the main requirement, and cassava or yams are a more convenient source of energy than a leaf crop. In arid, or semi-arid regions the case for LP is uncertain. Crops grown with irrigation, but in very dry air, tend to have rather dry, leathery leaves from which protein does not extract as easily as it does from lush leaves, and it is obviously easy to conserve fodder by drying when the air is dry. It is in humid regions where protein deficiency is the main nutritional problem that the case for substituting LP crops for some of the cereals, or even for protein-rich crops such as beans, is strongest. The evidence assembled in Chapter 3 shows that LP can give a greater yield of protein than any other crop. In regions where it rains nearly every day, so that it is difficult to ripen seed crops, the case is overwhelming. These are the regions where the first trials of family, or village, production of LP should be made. So many people live in such regions, that LP would make a contribution even if it were used nowhere else, and it is obviously wise to start at sites where the probability of success is greatest.

The potential advantages of finding out how to make LP, and how to use it in suitable regions, seem so obvious that outside observers are puzzled by the slow progress of work on the subject. I have discussed elsewhere the general problem of opposition to innovation – nutritional or otherwise (Pirie 1971*b*, 1972, 1975*b*). A specific obstacle is that thoughtful people in poor countries are understandably suspicious of the proposal that they should start using something that is not used elsewhere. The fact that climates and circumstances are different may not convince sceptics completely. Acceptance has not been helped by the tendency of various organisations, e.g. FAO (1964, 1969), United States President's Science Advisory Committee (1967), United Nations (1968), to accompany a phrase such as '... the potential value of such products is unquestionable ...' with a catalogue of all the possible difficulties that may be encountered with LP. Difficulties undoubtedly exist: it is unfortunate that it is only for LP that they are itemised although they obtrude to a similar extent with all the other novel protein sources. A foolish and tendentious statement by the Protein Advisory Group (1970) asserted that village production of LP was impractical. A more sensible statement from the same group was given only restricted circulation.

Commercial production

The gloomy conclusions come to by the organisations referred to in the last paragraph did not deter some animal feeding-stuffs manufacturers from taking an interest in the potentialities of fodder fractionation. Part of the reason for the failure of early attempts at commercialisation (p. 14) was the use of unsuitable equipment. This was probably also part of the reason for the ending of production by Batley-Janss (Brawley, California) and by Technion (Haifa). A contributory factor in these failures was the absence, at that time, of emotional commitment to energy conservation. France-Luzerne (Gastineau, 1974, 1976) started research on fodder fractionation in 1956 and started large-scale work at Chalons-sur-Marne and Mairy-sur-Marne in the 1970s. Daily production of dry LP had reached 2 t by 1974; an annual production of 1500 t is planned. France-Luzerne puts particular emphasis on the xanthophyll content of the LP, which is mainly used as feed for broilers, and on the fuel saved when the pressed fibre is dried. Dengie Crop Driers (Southminster, Essex) in cooperation with Simon-Barron Ltd (Gloucester) are beginning regular production of LP and dried fibre. Because of the cooperation of an engineering firm, improvements to the equipment is an important part of their program (Wilsdon, 1977; Freeman, 1977). In Wisconsin, Koegel & Bruhn (1977) are studying the potentialities of

L.P.F.F.—5**

a pulper, similar in design to a 'California' or 'pelleting press', in which the crop is forced through an array of holes 16 mm in diameter. They make some sensible comments on the importance of simplicity and cheapness and point out that industrial equipment tends to cost 10 times as much as agricultural equipment of the same weight. That was also my experience when several aircraft firms, wishing to diversify 15 years ago, produced designs for fractionation equipment.

Forage suitable for fractionation is most abundant in May, June, September and October; cereal straw is produced in August. Now that there is a revival of interest in increasing the digestibility of straw by treating it with alkali, and the clumsy techniques of soaking in dilute alkali are being replaced by treatment with small amounts of concentrated alkali, it is apparent that fodder fractionation and straw treatment dovetail neatly together. Many of the facilities of a factory can be used for both processes. BOCM-Silcock has a factory at Womersley in Yorkshire where the two processes are run in tandem. LP has been produced for many years in Hungary (Hollo & Koch, 1971); the present status of the project is uncertain. It is also difficult to get precise information about the level of commercial interest in Brazil, Canada, Eire, Japan, New Zealand, Norway, Philippines, Spain, Sweden, the USA and USSR. The situation is fluid in all these countries and there is some commercial secrecy.

The advisability of starting fodder fractionation in a region depends on the climate. If fine weather can be relied on in summer, so that partial field drying is reliable, and if the LP would be used only as broiler feed, there seems to be little advantage in extracting LP rather than separating the partly dried crop into protein-rich and protein-poor fractions mechanically (cf. Pirie 1977*a*). The strength of the arguments in favour of fodder fractionation in any region are proportional to the unpredictability of summer weather, and to the demand for conserved winter fodder.

Economic assessments

Various early estimates, or guesses, at the economics of fodder fractionation reflect the prejudices of the guesser rather than the reality of the situation. As experience accumulates it is becoming possible to coordinate all the various components of the probable economic picture. Dumont & Boyce (1976) developed a computer model which compared a traditional silage system with a system in which some of the grass on a livestock farm was fractionated. They set out clearly the assumptions they made, e.g. that 25 % of the DM and 40 % of the nitrogen in the grass would be extracted, that the 'whey' would be returned to the land, that two men could manage 80 cows and 750 pigs, and that the extra labour required to fractionate 1 t of DM in the original crop would be 2.8 man-hours. With these constraints, and with the amount of grass expected in an average year, the gross margin resulting from forage fractionation increased with increasing area up to 60 ha, and then remained constant at 2700 £ per year. To get that gross margin, the extra equipment needed to fractionate grass instead of making silage had to cost less than 11 130 £ and to be able to handle 5 to 10 t (wet weight) of crop per hour. According to this model, barley and concentrates in a year's feed for the pigs would be partly replaced by 46 t of LP. The authors stressed that they deliberately overestimated costs and the amount of labour needed, and made no allowance for the advantage of relative independence from weather when fodder is fractionated instead of being conserved conventionally.

Another fairly detailed study (Morris, 1977) approached the problem from the other end and, comparing LP with various conventional feeding stuffs as a source of protein, xanthophyll, carotene, energy, etc., concluded that it would be used in poultry feed if it could be produced for less than 170 £ per t. Yet another economic analysis (Wilkins, Heath, Roberts & Foxell, 1977) compared six ways in which fodder fractionation might be incorporated in a farming system. The largest item in the balance sheet was the cost of growing the crop: that cost is the same regardless of the manner in which the crop is used unless production depends for at least part of the year on a leafy by-product. The profitability of fractionation, compared with conventional dehydration, increased with increasing fuel costs and with diminishing equipment costs. Wilkins *et al.* limited their analysis to large units handling annually 10 000 t (DM) of lucerne, rather than fractionation on a farm of normal size in Britain; they concluded that the processes they studied were unlikely to be profitable.

In a general survey of the amount of energy needed to produce various types of protein concentrate, Slesser, Lewis & Edwardson (1977) list LP as one of the more economical. But there are still sceptics. A journalist (Anonymous, 1975) reported that economists at the Lord Rank Research Centre considered LP production impractical because the return on capital would be only 19 %! An Australian group expects to get a 30 % return on capital (Anonymous, 1976). The conclusion seems inescapable that, insofar as economics is a science at all, it is one of the experimental sciences. The profitability of fodder fractionation will not be known until it is tried in practical, sensible conditions. Even with such an experimental approach, the conclusion can be upset if there is a change in the complex system of taxation and subsidy arranged nationally and by organisations such as the European Economic Community. Physiological efficiency is no longer the only factor controlling profitability.

These points have not been overlooked by the bodies responsible for the direction of agricultural research in the UK. The report of the Energy Working Party (December 1974) of the Joint Consultative Organisation for Research and Development in Agriculture and Food comments: '. . . of particular interest in this connection are the prospects for extraction of protein from green crops by squeezing out part of the water mechanically which brings with it some of the proteinaceous dry matter.' And the Agricultural Development Advisory Service has set up a 'ways and means panel' to coordinate work on forage fractionation (Carter, 1977).

Discussion and economic assessment depend to an unreasonable extent on experience gained with current techniques. Some recapitulation is therefore worthwhile so as to clarify the possible effects of technical developments on the costs of the different processes in fodder fractionation.

(1) The costs of growing, harvesting and carting a crop are not affected by the use made of the crop except for the increased service that may be got from equipment because fodder fractionation depends less on weather than conventional conservation.

(2) It may be worthwhile using fertiliser and irrigation water at levels

that increase the yield of both protein and DM because, with fodder fractionation, the resulting fibre would not contain more protein than a ruminant could use effectively. This would give a better return on the capital value of the land.

(3) For the same annual production of dry fodder, a smaller drying unit will be needed if the fodder is fractionated and less fuel will be used. But fractionating equipment will have to be installed and it will use power. This is the crux of the discussion, and this is where information and technical development are most needed.

Sceptics and advocates make many valid but antithetical points when arguing about the amount of energy that could be saved by fractionation. Latent heat of vaporisation need not be wasted; if unfractionated fodder were dried in a 'vapour-compression' unit, liquid water would be discharged and much of the latent heat would be recovered. But the cost of such units is of a different order from the cost of conventional driers. The energy in the fuel used in a conventional drier is applied directly to the crop, whereas mechanical energy is used in fractionation; the conversion of the energy in fuel into electrical energy, or to mechanical energy in a diesel, is only 20 to 40 % efficient. On the other hand, in a conventional drier the transfer to a crop of the energy of combustion is far from 100 % efficient, and the energy 'wasted' by a diesel is discharged at a high enough temperature to be useful – for example, for drying the pressed fibre. Much more information is needed before the merits of fodder fractionation as a means of conserving energy can be accurately assessed. In the meantime we should heed the conclusion of an eminently practical paper by Bruhn, Straub & Koegel (1977) that the process '. . . may serve to delay the demise of the dehydration industry now operating on gas or liquid fuel'.

Alternative ways of using leaf protein as an animal feed

Other facets of the economic picture depend on assumptions made about the manner in which LP would be used. Feeding stuffs manufacturers assume that it must be separated and dried before it can be sold. This assumption is probably correct for them, because it is difficult to envisage a large-scale commercial distribution system that could cope with a moist, perishable press-cake. Several research institutes work on the assumption that forage fractionation would be a farm operation and that whole juice would be fed to pigs through a liquid-feed system. In principle, there are many advantages in such a system; it eliminates the need for filtration equipment, the cost of drying, and the damage to the LP if drying is mismanaged. Grass and lucerne juices, preserved in various ways (p. 68), were found by Maguire & Brookes (1972, 1973), Naumenko, Tarasenko & Kinsburgskii (1975), Naumenko, Morozova & Kinsburgskii (1977) and Houseman & Connell (1976) to replace conventional protein supplements as effectively as separated LP (p. 82). Lucerne juice was better than skim milk in calf diets (Prasad *et al.*, 1977). There have, however, been difficulties. Young pigs did not like lucerne juice (Connell, 1975), some old pigs did not thrive when half the fish meal in their feed was replaced by lucerne juice (Braude *et al.*, 1977), and calves did not like grass juice though pigs and chickens did well on it (Erkomaishvili, Chubinidze & Kitiashvili, 1976). Braude *et al.* (1977) suggest that there may be an excess of cations in the feed when juice is used as the sole protein supplement; there does not seem to be an excess of any individual ion, and extra water counteracted their combined effect.

In spite of the convenience of feeding juice straight from the extraction unit into the pig-feeding pipe line, I doubt the practicality of this arrangement. Depending on whether the crop is young or semi-mature, and whether it has dew or rain on it or is harvested at the end of a sunny afternoon, the protein content of the juice can vary by a factor of five, and its DM by a factor of three. This complicates the calculation of diets. Proteolysis in stored juice is probably innocuous because the products of hydrolysis have the same nutritive value as the original protein, but spontaneous coagulation may make juice distribution troublesome. Furthermore, much of the material in 'whey' is of doubtful nutritional value in nonruminants, and some components may be harmful. For all these reasons it is probably better to coagulate the juice, allow the coagulum to settle, and syphon off most of the 'whey'. The volume of the sediment depends on the protein content of the juice, but it has a fairly constant composition. When made by heat coagulation, 10 to 15 % of its wet weight is protein. This relative constancy makes the sediment a convenient component of liquid feeds. Because of the diminution in volume and the removal of some buffering material, it needs less acid (e.g. formic, phosphoric or propionic) or sulfite for preservation if that should be essential because of an uneven supply of forage.

The necessary improvements in extraction equipment

Whether the primary reason for fodder fractionation is small-scale production of LP for local use, with fibre as a by-product; or large-scale production of fibre that can be easily conserved, with juice or LP as a by-product, the equipment must be efficient, robust and simple. The computer model used by Dumont & Boyce (1976) set a tentative limit to its cost. Too little attention has been paid to the design of equipment for this precise job. Disintegration and pressing are in principle different processes: logically therefore, equipment suited to the one process would not be suited to the other. Because Rothamsted is primarily an agronomic institute, our early work was concerned more with demonstrating the feasibility of LP extraction and with measuring the possible yields from different crops and systems of husbandry, than with efficient methods for extraction in practice. For quantitative work it is essential that the conditions of extraction should be the same in each experiment: that is probably not feasible unless the crop is pulped and then pressed in equipment of the type described in Chapter 3. Larger units of the same type have been made (p. 12), and a great deal of LP has been produced in them. Because of their high speed they waste power by creating wind; more machines of this type are not likely to be made.

Extraction in one operation is obviously preferable when measurements are not being made, and all that is required is a supply of leaf juice and partly dewatered fibre. The three-roll mill gradually evolved during 400 years into a fairly efficient unit for crushing sugar cane stems: there was no reason to assume that it would be effective for extracting protein-rich juice from leaves. After being tried for that purpose in several institutes, it is now no longer used anywhere. Several institutes use various types of screw expeller. The principle of the screw expeller is admirable for work on well-lubricated materials in which the fluid that is to be expressed has already been released from cells or similar structures. It is for this reason that oil-seeds are usually cooked before pressing. A conventional screw expeller will not extract protein-rich leaf juice if it is truly efficient as a press: inefficiency in the application of pressure leads to some incidental rubbing which releases juice. Furthermore, most expellers exert excessive pressure. After pressing at as little as 2 kg per cm² (200 kPa) for a few seconds on a grid, the DM of a layer of leaf pulp 10 mm thick increases to more than 30 %.

In the late 1940s a large cordite mixer (Pfleiderer) was used at Rothamsted to produce batches of pulp with which presses could be tested. This demonstrated that slow rubbing was an effective method for liberating protein-rich juice from leaves. It should be possible to design a unit in which the crop is effectively rubbed and disintegrated while it is being pushed into the section of an expeller in which juice will be pressed out, and in which pressure is applied in a manner which combines further rubbing with some rearrangement. In such a unit, little energy would be wasted in friction between a smooth scroll and a mass of largely undamaged leaf. The primary objective in a standard screw expeller is to apply pressure – rubbing is inci-dental, though there is more rubbing when two screws work together. Liberation of juice should be the first step in an efficient extraction unit – there is little advantage in pressing until that has been done.

A unit designed with these principles in mind (Pirie 1977*b*) consists of a perforated cylindrical barrel, with a slow-moving, conical rotor in it carrying an interrupted flight, or set of flights. The crop is fed in at the narrow end of the cone by an auger; this also is interrupted so that the crop is extensively disintegrated before it enters the barrel. Because crops vary greatly in texture and DM content, arrangements are made to vary the feeding rate of the auger and the extent to which the cone is inserted into the barrel, i.e. the thickness of the emerging shell of pressed fibre. More detailed description is unnecessary because the unit is still being improved.

The experimental unit takes 1 to 2 kg (fresh weight) of crop per min and runs at about 300 watts. Larger units with the same general appearance could be made. A unit to take several tonnes an hour could operate on the same principle, but it would probably be better to alter its appearance radically. Instead of a narrow-angle cone rotating about a horizontal axis, it should have a wide-angle, stationary, perforated cone with its axis vertical and its apex pointing downwards. The cylinder would be replaced by a rotating annulus, pressed down on to the edges of the cone by a set of springs. This arrangement would be more flexible and less vulnerable to stones, which become a serious hazard in large-scale work.

In spite of the advantages of doing the whole job in one operation in one machine, it seems likely that energy is wasted when unlubricated material is pressed heavily in a machine with the general character of a screw expeller so as to produce fibre containing 30 to 40 % DM. As already mentioned (p. 19), juice coming from pulp while pressure is being applied differs in composition from juice coming out later. Because the leaf fibres pack together and form a fine filter, the early fraction of juice contains most of the LP. Later fractions may contain so little that they hardly justify the cost of the steam used for coagulation. When silage will be made, there is no advantage in pressing out as much juice as possible, but heavy pressing is advantageous when the fibre will be dried. From suitably pretreated material, in the moisture range with which we are here concerned, juice can be expressed for one-thousandth the expenditure of energy that would have been needed to evaporate the same amount of water. This assumes that, as in conventional grass-drying, latent heat is not being recovered.

In belt presses of the type used for making LP (p. 13), energy is not wasted in friction to the same extent as it is wasted in a screw expeller. However, they cannot apply the necessary pressure because of the limited strength of flexible belts. But the basic principle of a belt press, in which there is no relative movement between compressed fibre and the perforated surface, is essential for economy. Rollers can apply sufficient pressure but it is maintained so briefly that the expressed juice has not time to run away. For a time we (Davys & Pirie, 1960) used a press with intermittent action, but it was clumsy. Secondary pressing could probably be most efficiently managed in a press of the 'horn angle' type. In this, the material to be pressed is fed into the space between a drum and a perforated cylinder, of slightly greater diameter, which is pressed against the drum. Presses such as this maintain pressure for much longer than roller presses of similar dimensions and rotating at the same rate. So far as I know, presses of this type have not yet been used for fodder fractionation.

Most of those who have thought about the design of machinery for fodder fractionation wonder at some stage whether there would be any advantage in making the machinery mobile. The advantage is obvious when water weeds are being processed (p. 38): there could be ample space on a barge, and the transfer of harvested weed to a land-based unit would be troublesome. I can think of no other circumstances in which it would be easier to move the machinery to the crop rather than the crop to the machinery.

Crops vary greatly in texture and water content, and different amounts of juice are expressed by the machinery used in different institutes where fodder fractionation is studied. The figures for power consumption that appear in annual reports, and in papers given at symposia, are therefore not comparable: they range from 6 to 15 kW h per t (wet weight) of crop. Improvements in the extraction equipment used at Rothamsted diminished power consumption from about 30 to about 10 kW h per t between 1955 and 1970. When equipment has been designed that does not waste energy by creating wind, by useless friction, and by applying excessive pressure, it should be possible to use only 2 kW h for processing 1 t of crop.

The technique of fodder fractionation is in its infancy. Economic assessment has been attempted on the basis of existing crops and existing equipment, with insufficient attention to the effects of selecting crops and designing equipment specifically for this purpose. Textbooks of biochemical engineering deal almost exclusively with microbial conversions and neglect the less wasteful process of separation. Conversions are often necessary, but separation will sometimes suffice. By making optimal use of all the products of a fractionation, processes that seem questionable in isolation can become viable. Hartley (1937) saw this clearly 40 years ago: 'Finally there is the social aspect of the problem. Hitherto agriculture and industry have represented divergent interests, the town and the village, the factory and the farm, the huge closely knit corporations wanting cheap food and raw materials, contrasted with the scattered, unorganised agricultural producers at the mercy of nature for their output. The modern development of factories in close touch with farmers in the processing of their goods should bring industry and agriculture closer together and help to establish a community of interest between them. What their joint efforts can accomplish will depend on the progress of knowledge in biochemistry and chemistry on the one hand, and in plant breeding and cultivation on the other. With the modern technique of genetics and the closer association of the farmer and the manufacturer, there is a fascinating prospect of new strains of plants that will yield the ideal products for industry almost to a standard specification. Then indeed we should have realised Bacon's ideal of commanding nature in action.'

Appendix:
Hilaire Marin Rouelle
(1718–79)

Rouelle's paper is now often referred to, but seems seldom to have been read. A free translation may therefore be useful. In modern French, *fécule* means starch. Rouelle used the word in more than one sense. It was possibly because of this that Fourcroy doubted his conclusions and suggested that Rouelle had simply ground the leaf to such a fine pulp that it passed through the cloth used for straining. This suggestion incensed Proust – the discoverer of leucine and an early contributor to our knowledge of citric acid. Proust (1803) wrote of the '. . . beautiful liquid velvet expressed from leaves', but he used *fécule* in as imprecise a manner as Rouelle. *Fécule*, and the various words for weights and volumes, are best left untranslated.

Observations on the *Fécules* or green parts of Plants, and on glutinous or vegeto-animal matter; by M. Rouelle, demonstrator in chemistry in the 'Jardin du Roi'

The *fécules* or green parts of plants were classed as resins by my late brother because of their solubility in all oily solvents and in alcohol. He defined these *fécules* as being composed of (1) green resinous colouring matter and (2) parenchyma or plant fibres, separated by pounding with a pestle; and he pointed out that when either the *fécules* or the plant juices were heated with oils or fats that dissolve the green part, some insoluble matter always remained which he thought, as I have just said, belonged to to the parenchyma or fibrous portion of the plant.

I have given, in the *Journal de Médecine* of last March and in the *Avant Coureur*, etc., analyses of several *fécules* or green parts of plants. I have explained that *fécules* made from different plant families, after being dried, give, when analysed in a retort, the same products as animal substances; this proves that the *fécules* or green-coloured parts of plants are not made of pure vegetable matter, because the analytical products of vegetable matter are not found in them, but on the contrary, those of animal matter.

When I wrote of this in the *Journal de Médecine* and when I said that 'the green *fécules* were not resins because the products of their analysis were quite different from those of all known resins', I did not think I needed to give a clearer explanation of the nature of this substance; I stated, however, that the presence in all vegetable matter of a substance completely similar to the glutinous matter of wheat could be demonstrated. I laid aside for a further paper the more precise description of this kind of substance, which is in fact composed of a pure resin which gives the green colour to all plants, and of this glutinous or vegeto-animal matter.

The glutinous matter, which is found in all the *fécules*, is usually more abundant than the resinous green part, the latter making up, in the plants that I have examined, only a fifth to a third of the *fécule*. I will confine myself for the moment to a single example, and indicate some others.

Fécule *of hemlock*

The required amount of hemlock, when it is almost in flower, is ground carefully in a marble mortar with a wooden pestle. The juice is strained through a well-stretched cloth, or a cloth filter, and heated to the point at which one can hold a finger in it for several minutes. The *fécule* separates out and part of it floats on top of the liquid; part sinks or remains suspended. The whole is put on a cloth filter, the liquid becomes clear; the sediment remains on the cloth and is carefully collected. This is the procedure usually followed to make these plant *fécules*.

Remarks

(I) By paying a little attention to what goes on in this process, it is clear that the greenest part of the *fécule* separates first. When the heat is increased, distinct and well-defined flakes or white dots appear below the *fécule* which floats on the surface. These observations suggest that there are two *fécules*.

(II) If juice is heated to the temperature of milk when drawn from a cow and then taken off the fire, the *fécule* which separates has a more beautiful green colour than the one in Remark 1. If it is immediately poured onto a cloth, the filtrate is still slightly green.

(III) If liquid separated as above is heated more strongly, a *fécule* separates which is slightly green. This second *fécule* contains more glutinous, or vegeto-animal matter, than the first although, as will be seen, that contains an abundance.

(IV) *Fécule* made in the usual way is put into earthenware or glazed vessels and diluted, using a wooden stirrer, by pouring into each vessel 8 or 9 *pintes* of water. After standing for 24 hours to allow the *fécule* to sediment, the water is decanted; this is repeated a second and third time, leaving it to stand for 24 hours each time. After the third wash, the *fécule* is put on a cloth mounted on a wooden frame so as to get out as much moisture as possible; the *fécule* and cloth are then put on a plaster tile to absorb most of the remaining water. The *fécule* becomes fairly solid, it is cut into little pieces and put to dry on paper placed on sieves.

(V) Several methods can be used to separate the two *fécules*; but some are difficult and others expensive. The easiest uses alcohol, which has no action on vegeto-animal matter but dissolves the green colouring matter. The dried *fécule* has only to be ground to a fine powder and digested with alcohol several times.

The alcohol dissolves the green part and leaves the vegeto-animal part; but the process takes a very long time because of the way the two substances adhere during drying; I have found that it is difficult to separate them completely.

(VI) The two substances are more easily separated if the above procedure is followed using fresh *fécule*. When it is ready to be dried, it is mixed carefully with alcohol, digested for 24 hours in a water bath or sand bath, and left to cool. The alcohol is then decanted or filtered. This digestion is repeated a second and third time. The three extracts are put into the tin boiler of a distilling apparatus; the alcohol passes over as a clear liquid which smells of hemlock. A soft, resinous substance, which clings to the fingers like terebinth resin, remains in the boiler. The alcohol that has been distilled off is used to re-extract the *fécule* and the process is repeated until the *fécule* no longer gives up any green colour. The quantity of green colouring matter varies a little; it depends first on the state and age of the plant, and secondly on whether the leaves or the stalks are taken.

(VII) The *fécule* remaining is a dirty greyish-white, and blackens on drying. It alone makes up three-quarters of the amount of matter used in the experiment; this is the glutinous or vegeto-animal substance, as its analysis will show.

(VIII) When 4 *onces* of this substance were slowly heated in a retort in a reverberatory furnace, a small amount of distillate and drops of volatile alkali came over first. When the heat was increased, the alkali became more concentrated, and finally became solid. At the same time, an oil passed over which floated on the alkaline fluid. This resembles the oil obtained from the vegeto-animal matter of flour and from the caseous part of milk.

The residue is fairly bulky. It has a fairly even texture because the pieces of the glutinous substance have softened and stuck together. It closely resembles the residue from the glutinous part of corn and the caseous part of milk. It weighs an *once* or more.

(IX) A few drops of distillate passed over when 2 *onces* of the green colouring matter were heated gently in a glass retort. When the heat was increased, an acid liquid came over and increased in strength, then a beautiful yellow oil which became darker as it thickened. The acid is very strong and resembles that obtained from wax. The oil is light and floats on the acid, as do most oils obtained from resins. Finally, the residue is fairly bulky and light and weighs 3 *gros* and 60 *grains*.

(X) Rosemary, from which the extractable part has been removed by repeated decoctions, is only, according to Boerhaave, the dross or skeleton of the plant. It still contains a small amount of crude oil which gives a little flame, but its ash does not give fixed alkali.

This learned doctor did not know that this exhausted rosemary gives fixed alkali and contains a green colouring matter which is soluble in fats, oils, resins and alcohol.

My brother, as I have already said, and as we have written – it was indeed printed long ago, and is in short known by everybody – was the first to demonstrate this green part in exhausted rosemary; but there are also in the rosemary left after extraction with water and alcohol, two substances, namely: (1) a very small quantity of glutinous or vegeto-animal matter, and (2) a substance that has a vegetable nature, insoluble in water and alcohol. The rosemary left after extraction with these two solvents still gives to a fairly marked degree some acid and some oil when distilled. Consequently, there remains in the rosemary a constituent or a substance which was hitherto unknown and which escapes the action of these two solvents. Later, I will describe this substance more fully.

References

Akeson, W. R. & Stahmann, M. A. (1965). Nutritive value of leaf protein concentrate, an *in vitro* digestion study. *J. agric. Fd Chem.*, **13**, 145.

Akinrele, I. A. (1963). The manufacture and utilisation of leaf protein. *J. W. Afr. Sci. Ass.*, **8**, 74.

Allen, G. (1976). Some aspects of planning world food supplies. *J. agric. Econ.*, **27**, 97.

Allison, F. E. (1973). *Soil organic matter and its role in crop production.* Elsevier, Amsterdam.

Allison, R. M. (1971). Factors influencing the availability of lysine in leaf protein. In *Leaf protein: its agronomy, preparation, quality and use*, ed. N. W. Pirie, p. 78. Blackwell, Oxford.

Allison, R. M., Laird, W. M. & Synge, R. L. M. (1973). Notes on a deamination method proposed for determining 'chemically available lysine' of proteins. *Br. J. Nutr.*, **29**, 51.

Allison, R. M. & Vartha, E. W. (1973). Yields of protein extracted from irrigated lucerne. *N.Z. J. exp. Agric.*, **1**, 35.

Anderson, J. W. & Rowan, K. S. (1967). Extraction of soluble leaf enzymes with thiols and other reducing agents. *Phytochemistry*, **6**, 1047.

Anonymous (1965). *Investigations into the production of high protein concentrate from leaves for inclusion in the diet of infants and children.* Sci. Res. Coun. Tech. Rep., Kingston, Jamaica.

Anonymous (1974). Alfalfa protein for human use. *Agric. Sci. Rev.*, **11**(2), 55.

Anonymous (1975). Accountants won't swallow leaf protein. *New Scient.*, **66**, 388.

Anonymous (1976). Making the most out of lucerne. *Rural Res. C.S.I.R.O.* **91**, 10.

Arkcoll, D. B. (1969). Preservation of leaf protein by air drying. *J. Sci. Fd Agric.*, **20**, 600.

Arkcoll, D. B. (1971). Agronomic aspects of leaf protein production in Great Britain. In *Leaf protein: its agronomy, preparation, quality and use*, ed. N. W. Pirie, p. 9. Blackwell, Oxford.

Arkcoll, D. B. (1973*a*). The preservation and storage of leaf protein preparations. *J. Sci. Fd Agric.*, **24**, 437.

Arkcoll, D. B. (1973*b*). Effects of leaf protein 'whey' on soil. *A. Rep. Rothamsted exp. Stn*, 1972, 117.

Arkcoll, D. B. & Festenstein, G. N. (1971). A preliminary study of the agronomic factors affecting the yield of extractable leaf protein. *J. Sci. Fd Agric.*, **22**, 49.

Arkcoll, D. B. & Holden, M. (1973). Changes in chloroplast pigments during the preparation of leaf protein. *J. Sci. Fd Agric.*, **24**, 1217.

Bagchi, D. K. & Matai, S. (1976). Studies on the performance of tetrakali (*Phaseolus aureus* Linn.) as a leaf protein yielding crop in West Bengal. *J. Sci. Fd Agric.*, **27**, 1.

Balasundaram, C. S., Krishnamoorthy, K. K., Chandramani, R., Balakrishnan, T. & Ramadoss, C. (1974*a*). The yield of leaf protein extracted by large scale processing of various crops. *Indian J. Home Sci.*, **8**, 6.

Balasundaram, C. S., Krishnamoorthy, K. K., Balakrishnan, T., Ramadoss, C. & Chandramani, R. (1974*b*). Screening plant species for leaf protein extraction. *Indian J. Home Sci.*, **8**, 1.

Balasundaram, C. S., Chandramani, R., Krishnamoorthy, K. K. & Balakrishnan, T. (1975). Optimum time of cutting for maximum yield of extractable protein from some fodder grasses. *Madras agric. J.*, **62**, 431.

Balasundaram, C. S. & Samuel, D. M. (1968). Incorporation of different leaf proteins in South Indian dishes. *Madras agric. J.*, **55**, 540.

Barber, R. S., Braude, R. & Mitchell, K. G. (1959). Leaf protein in rations of growing pigs. *Proc. Nutr. Soc.*, **18**, iii.

Barnes, M. F. (1976). Growth of yeasts on spent lucerne whey and their effectiveness in scavenging residual protein. *N.Z. J. agric. Res.*, **19**, 537.

Bates, R. P. & Hentges, J. F. (1976). Aquatic weeds – eradicate or cultivate. *Econ. Bot.*, **30**, 39.

Batra, U. R., Deshmukh, M. G. & Joshi, R. N. (1976). Factors affecting extractability of protein from green plants. *Indian J. Pl. Physiol.*, **19**, 211.

Bawden, F. C. & Kleczkowski, A. (1945). Protein precipitation and virus inactivation by extracts of strawberry plants. *J. Pomol.*, **21**, 2.

Bawden, F. C. & Pirie, N. W. (1938). A note on some protein constituents of normal tobacco and tomato leaves. *Br. J. exp. Path.*, **19**, 264.

Bawden, F. C. & Pirie, N. W. (1944). The liberation of virus together with materials that inhibit its precipitation with antiserum, from the solid leaf residues of tomato plants suffering from bushy stunt. *Br. J. exp. Path.*, **25**, 68.

Ben Aziz, A., Grossman, S., Budowski, P., Ascarelli, I. & Bondi, A. (1968). Antioxidant properties of lucerne extracts. *J. Sci. Fd Agric.*, **19**, 605.

Ben Aziz, A., Grossman, S., Budowski, P. & Ascarelli, I. (1971). Enzymic oxidation of carotene and linoleate by alfalfa: properties of active fractions. *Phytochemistry*, **10**, 1823.

Betschart, A. A. (1977). The incorporation of leaf protein concentrates and isolates in human diets. In *Green crop fractionation*, ed. R. J. Wilkins, p. 83. Occasional Symposia 9, Br. Grassld Soc., Maidenhead.

Betschart, A. A. & Kinsella, J. E. (1974a). Influence of storage on composition, amino acid content and solubility of soybean leaf protein concentrate, *J. agric. Fd Chem.*, **22**, 116.

Betschart, A. A. & Kinsella, J. E. (1974b). Influence of storage on *in vitro* digestibility of soybean leaf protein concentrate. *J. agric. Fd Chem.*, **22**, 672.

Bickoff, E. M., Bevenue, A. & Williams, K. T. (1947). Alfalfa has a promising chemurgic future. *Chemurg. Dig.*, **6**, 215.

Bickoff, E. M. & Kohler, G. O. (1974). Preparation of edible protein of leafy green crops such as alfalfa. *US Pat.* 3 823 128.

Bickoff, E. M., Booth, A. N., de Fremery, D., Edwards, R. H., Knuckles, B. E., Miller, R. E., Saunders, R. M. & Kohler, G. O. (1975). Nutritional evaluation of alfalfa leaf protein concentrate. In *Protein nutritional quality of foods and feeds*, ed. M. Friedman, p. 319. Dekker, New York.

Bosshard, H. (1972). Über die Anlagerung von Thioäthern an Chinone und Chinonimine in stark sauren Medien. *Helv. chim. Acta*, **55**, 32.

Boyd, C. E. (1968). Fresh-water plants: a potential source of protein. *Econ. Bot.*, **22**, 359.

Boyd, C. E. (1971). Leaf protein from aquatic plants. In *Leaf protein: its agronomy, preparation, quality and use*, ed. N. W. Pirie, p. 44. Blackwell, Oxford.

Boyd, C. E. (1976). Accumulation of dry matter, nitrogen and phosphorus by cultivated water hyacinth. *Econ. Bot.*, **30**, 51.

Braude, R., Jones, A. S. & Houseman, R. A. (1977). The utilization of the juice extracted from green crops. In *Green crop fractionation*, ed. R. J. Wilkins, p. 47. Occasional Symposia 9, Br. Grassld Soc., Maidenhead.

Bray, W. (1976). Leaf-protein concentrate as a source of vitamins. *Proc. Nutr. Soc.*, **35**, 6A.

Bray, W. (1977). The separation and preservation of leaf protein concentrates and isolates for use as human food. In *Green crop fractionation*, ed. R. J. Wilkins, p. 107. Occasional Symposia 9, Br. Grassld Soc., Maidenhead.

Bray, W. J., Humphries, C. & Ineritei, M. S. (1978). The use of solvents to decolourise leaf protein concentrate. *J. Sci. Fd Agric.*, **29**, 165.

Bris, E. J., Hibbits, A. G., Algeo, J. W., Wooden, G. R. & Batley, B. (1970). A comparison of rations with concentrated alfalfa extract and cane molasses as to volatile fatty acid production *in vitro* and digestibility *in vivo*. *J. Anim. Sci.*, **31**, 237.

Brown, H. E., Stein, E. R. & Saldana, G. (1975). Evaluation of *Brassica carinata* as a source of plant protein. *J. agric. Fd Chem.*, **23**, 545.

Bruhn, H. D., Straub, R. J. & Koegel, R. G. (1977). On farm forage protein – the potential and the means. Paper 3 at the A. Conf. Inst. agric. Engineers. London.

Bryant, M. & Fowden, L. (1959). Protein composition in relation to age of daffodil leaves. *Ann. Bot.*, **23**, 65.

Buchanan, A. R. (1969a). *In vivo* and *in vitro* methods of measuring nutritive value of leaf protein preparations. *Br. J. Nutr.*, **23**, 533.

Buchanan, A. R. (1969b). Effect of storage and lipid extraction on the properties of leaf protein. *J. Sci. Fd Agric.*, **20**, 359.

Buchanan, A. R. & Byers, M. (1969). Interference by cyanide with the measurements of papain hydrolysis. *J. Sci. Fd Agric.*, **20**, 364.

Butt, A. M., Ahmad, S. U. & Shah, F. H. (1972). Propagation of yeast on leaf protein concentrate by-products. *Pakist. J. scient. ind. Res.*, **15**, 208.

Byers, M. (1961). The extraction of protein from leaves of some plants growing in Ghana. *J. Sci. Fd Agric.*, **12**, 20.

Byers, M. (1967a). The *in vitro* hydrolysis of leaf proteins. I. The action of papain on protein extracted from the leaves of *Zea mays*. *J. Sci. Fd Agric.*, **18**, 28.

Byers, M. (1967b). II. The action of papain on protein concentrates extracted from leaves of different species. *J. Sci. Fd Agric.*, **18**, 33.

Byers, M. (1971a). The amino acid composition of some leaf protein preparations. In *Leaf protein: its agronomy, preparation, quality and use*, ed. N. W. Pirie, p. 95. Blackwell, Oxford.

Byers, M. (1971b). The amino acid composition and *in vitro* digestibility of some protein fractions from three species of leaves of various ages. *J. Sci. Fd Agric.*, **22**, 242.

Byers, M. (1975). Relationship between total N, total S and the S-containing amino acids in extracted leaf protein. *J. Sci. Fd Agric.*, **27**, 135.

Byers, M. & Jenkins, G. (1961). Effect of gibberellic acid on the extraction of protein from the leaves of spring vetches (*Vicia sativa* L.). *J. Sci. Fd Agric.*, **12**, 656.

Byers, M., Green, S. H. & Pirie, N. W. (1965). The presentation of leaf protein on the table II. *Nutrition*, **19**, 63.

Byers, M. & Sturrock, J. W. (1965). The yields of leaf protein extracted by large-scale processing of various crops. *J. Sci. Fd Agric.*, **16**, 341.

Carlsson, R. (1975). Selection of centrospermae and other species for production of leaf protein concentrates. Ph.D. thesis, University of London. (Many copies of this thesis are in circulation.)

Carpenter, K. J., Duckworth, J. & Ellinger, G. M. (1952). The supplementary protein value of a by-product from grass processing. *Br. J. Nutr.*, **6**, xii.

Carr, J. R. & Pearson, G. (1974). Nutritive values of lucerne leaf-protein concentrate and lupin-seed meal as protein supplements to barley diets for growing pigs. *N.Z. Soc. Anim. Prod.*, **34**, 95.

Carr, J. R. & Pearson, G. (1976). Photosensitisation, growth performance and carcass measurements of pigs fed diets containing commercially prepared lucerne leaf-protein concentrate. *N.Z. J. exp. Agric.*, **4**, 45.

Carruthers, I. B. & Pirie, N. W. (1975). The yields of extracted protein, and of residual fibre, from potato haulm taken as a by-product. *Biotechnol. Bioeng.*, **17**, 1775.

Carter, E. S. (1977). Forage fractionation development programme. *ARC News*, March 1977, p. 4.

Casselman, T. W., Green, V. E., Allen, R. J. & Thomas, F. H. (1965). *Mechanical dewatering of forage crops. Agric. exp. Stn Univ. Florida Tech. Bull.*, **694**.

Chalmers, M. I. & Synge, R. L. M. (1954). The digestion of protein and nitrogenous compounds in ruminants. *Adv. Protein Chem.*, **9**, 93.

Chan, P. H., Sakano, K., Singh, S. & Wildman, S. G. (1972). Crystalline fraction 1 protein: preparation in large yield. *Science*, **176**, 1145.

Chandramani, R., Balasundaram, C. S., Krishnamoorthy, K. K. & Balakrishnan, T. (1975a). Effect of nitrogen and the frequency of cutting in guinea grass (*Panicum maximum* L.). *Madras agric. J.*, **62**, 155.

Chandramani, R., Krishnamoorthy, K. K., Balasundaram, C. S. & Balakrishnan, T. (1975b). Optimum time of cutting for obtaining maximum yield of extractable protein from fenugreek (*Trigonella foenum-graecum*) varieties. *Madras agric. J.*, **62**, 230.

Chayen, I. H. (1959). Protein production by the impulse process. *Engineering, London*, **188**, 307.

Chayen, I. H., Smith, R. S., Tristram, G. R., Thirkell, D. & Webb, T.

(1961). The isolation of leaf components. *J. Sci. Fd Agric.*, **12**, 502.

Cheeke, P. R. (1974). Nutritional evaluation of alfalfa protein concentrate with rats, swine and rabbits. In *Proc. 12th tech. alfalfa conf. USDA*, p. 76. USDA, Washington.

Cheeke, P. R. & Garman, G. R. (1974). Influence of dietary protein and sulfur amino acid levels on the toxicity of *Senecio jacobaea* (Tansy ragwort) in rats. *Nutr. Rep. Int.*, **9**, 193.

Cheeke, P. R., Kinzell, J. H., de Fremery, D. & Kohler, G. O. (1977). Freezedried and commercially-prepared alfalfa protein concentrate evaluation with rats and swine. *J. Anim. Sci.*, **44**, 772.

Chen, I. & Mitchell, H. L. (1973.) Trypsin inhibitors in plants. *Phytochemistry*, **12**, 327.

Chibnall, A. C. (1939). *Protein metabolism in the plant*. Yale Univ. Press, New Haven.

Chibnall, A. C., Rees, M. W. & Lugg, J. W. H. (1963). The amino acid composition of leaf proteins. *J. Sci. Fd Agric.*, **14**, 234.

Clare, N. T. (1952). Photosensitization in diseases of domestic animals. *Commonw. Bur. Anim. Hlth Rev. Ser.*, **3**.

Clarke, E. M. W. & Ellinger, G. M. (1967). Fractionation of plant material. 2. Amino acid composition of some fractions obtained from broad bean plant (*Vicia faba* L.) and chromatographic differentiation of hydroxyproline isomers. *J. Sci. Fd Agric.*, **18**, 536.

Cohen, M., Ginoza, W., Dorner, R. W., Hudson, W. R. & Wildman, S. G. (1956). Solubility and color characteristics of leaf proteins prepared in air and nitrogen. *Science*, **124**, 1081.

Colker, D. A., Eskew, R. K. & Aceto, N. C. (1948). Preparation of vegetable leaf meals. *USDA Tech. Bull.*, **958**, 53.

Connell, J. (1975). The prospects for green crop fractionation. *Span*, **18**, 103.

Connell, J. & Foxell, P. R. (1976). Green crop fractionation, the products and their utilization by cattle, pigs and poultry. *Bienn. Rev. natn. Inst. Res. Dairying*, 1976, 21.

Connell, J. & Houseman, R. A. (1977). The utilisation by ruminants of the pressed green crops from fractionation machinery. In *Green crop fractionation*, ed. R. J. Wilkins, p. 57. Occasional Symposia 9, Br. Grassld Soc., Maidenhead.

Cowey, C. B., Pope, J. A., Adron, J. W. & Blair, A. (1971). Studies on the nutrition of marine flatfish. Growth of the plaice (*Pleuronectes platessa*) on diets containing proteins derived from plants and other sources. *Mar. Biol.*, **10**, 145.

Cowlishaw, S. J., Eyles, D. E., Raymond, W. F. & Tilley, J. M. A. (1956a). Nutritive value of leaf protein concentrates. I. Effect of addition of cholesterol and amino-acids. *J. Sci. Fd Agric.*, **7**, 768.

Cowlishaw, S. J., Eyles, D. E., Raymond, W. F. & Tilley, J. M. A. (1956b). II. Effects of processing methods. *J. Sci. Fd Agric.*, **7**, 775.

Crook, E. M. (1946). The extraction of nitrogenous materials from green leaves. *Biochem. J.*, **40**, 197.

Crook, E. M. & Holden, M. (1948). Some factors affecting the extraction of nitrogenous materials from leaves of various species. *Biochem. J.*, **43**, 181.

Darwin, C. (1868). *Variation of plants and animals under domestication.* Murray, London.

Davies, M., Evans, W. C. & Parr, W. H. (1952). Biological values and digestibilities of some grasses, and protein preparations from young and mature species, by the Thomas–Mitchell method, using rats. *Biochem. J.* **52**, xxiii.

Davies, R., Laird, W. M. & Synge, R. L. M. (1975). Hydrogenation as an approach to study of reactions of oxidizing polyphenols with plant proteins. *Phytochemistry*, **14**, 1591.

Davies, W. L. (1926). The proteins of green forage plants. 1. The proteins of some leguminous plants. *J. agric. Sci. Camb.*, **16**, 280.

Davys, M. N. G. & Pirie, N. W. (1960). Protein from leaves by bulk extraction. *Engineering, London*, **190**, 274.

Davys, M. N. G. & Pirie, N. W. (1963). Batch production of protein from leaves. *J. agric. Engng Res.*, **8**, 70.

Davys, M. N. G. & Pirie, N. W. (1965). A belt press for separating juices from fibrous pulps. *J. agric. Engng Res.*, **10**, 142.

Davys, M. N. G. & Pirie, N. W. (1969). A laboratory-scale pulper for leafy plant material. *Biotechnol. Bioeng.*, **11**, 517.

Davys, M. N. G., Pirie, N. W. & Street, G. (1969). A laboratory-scale press for extracting juice from leaf pulp. *Biotechnol. Bioeng.*, **11**, 528.

de Fremery, D., Bickoff, E. M. & Kohler, G. O. (1972). PRO-XAN process: air drying of alfalfa leaf protein concentrate. *J. agric. Fd Chem.*, **20**, 1155.

de Fremery, D., Miller, R. E., Edwards, R. H., Knuckles, B. E., Bickoff, E. M. & Kohler, G. O. (1973). Centrifugal separation of white and green protein fractions from alfalfa juice following controlled heating. *J. agric. Fd Chem.*, **21**, 866.

de Fremery, D., Edwards, R. H., Miller, R. E., Knuckles, B. E., Bickoff, E. M. & Kohler, G. O. (1974). Composition and uses of PRO-XAN and pressed residue. In *Proc. 12th tech. alfalfa conf. USDA*, p. 73. USDA, Washington.

Derban, L. K. A. (1975). Some environmental health problems associated with industrial development in Ghana. In *Health and industrial growth*, ed. K. Elliott & J. Knight, p. 49. Ciba Foundation Symposium, Elsevier, Amsterdam.

Derbyshire, J. C., Gordon, C. H., Holdren, R. D. & Menear, J. R. (1969). Evaluation of dewatering and wilting as moisture reduction methods for hay-crop silage. *Agron. J.*, **61**, 928.

Deshmukh, M. G., Gore, S. B., Mungikar, A. M. & Joshi, R. N. (1974). The yields of leaf protein from various short-duration crops. *J. Sci. Fd Agric.*, **25**, 717.

Deshmukh, M. G. & Joshi, R. N. (1969). Leaf protein from some leguminous plants. *Sci. Cult.*, **35**, 629.

Deshmukh, M. G. & Joshi, R. N. (1973). Effect of rhizobial inoculation on the extraction of protein from the leaves of cowpea (*Vigna sinensis* L. Savi ex Hassk.). *Indian J. agric. Sci.*, **43**, 539.

Dev, D. V., Batra, U. R. & Joshi, R. N. (1974). The yields of extracted leaf protein from lucerne (*Medicago sativa* L.). *J. Sci. Fd Agric.*, **25**, 725.

Dev, D. V. & Joshi, R. N. (1969). Extraction of protein from some plants of Aurangabad. *J. biol. Sci.*, **12**, 15.

Devi, A. V., Rao, N. A. N. & Vijayaraghavan, P. K. (1965). Isolation and composition of leaf protein from certain species of Indian flora. *J. Sci. Fd Agric.*, **16**, 116.

Doraiswamy, T. R., Singh, N. & Daniel, V. A. (1969). Effects of supplementing ragi (*Eleusine coracana*) diets with lysine or leaf protein on the growth and nitrogen metabolism of children. *Br. J. Nutr.*, **23**, 737.

Duckworth, J. & Woodham, A. A. (1961). Leaf protein concentrates. I. Effect of source of raw material and method of drying on protein value for chicks and rats. *J. Sci. Fd Agric.*, **12**, 5.

Duckworth, J., Hepburn, W. R. & Woodham, A. A. (1961). Leaf protein concentrates. II. The value of a commercially dried product for newly-weaned pigs. *J. Sci. Fd Agric.*, **12**, 16.

Dumont, A. G. & Boyce, D. S. (1976). Leaf protein production and use on the farm: an economic study. *J. Br. Grassld Soc.*, **31**, 153.

Dutrow, G. F. (1971). *Economic implications of silage sycamore. USDA Forest Serv. Res. Pap.*, SO **66**.

Dye, M., Medlock, O. C. & Christ, J. W. (1927). The association of vitamin A with greenness in plant tissue. I. The relative vitamin A content of head and leaf lettuce. *J. biol. Chem.*, **74**, 95.

Dykyjova, D. (1971). Productivity and solar energy conversion in reedswamp stands in comparison with outdoor mass cultures of algae in the temperate climate of central Europe. *Photosynthetica*, **5**, 329.

Edwards, R. H., Miller, R. E., de Fremery, D., Knuckles, B. E., Bickoff, E. M. & Kohler, G. O. (1975). Pilot plant production of an edible white fraction leaf protein concentrate from alfalfa. *J. agric. Fd Chem.*, **23**, 620.

Eggum, B. O. & Christensen, K. D. (1975). Influence of tannin on protein utilization in feedstuffs with special reference to barley. In *Breeding for seed protein improvement*, p. 135. International Atomic Energy Agency, Vienna.

Eichenberger, W. & Grob, E. C. (1965). Beiträge zur Chemie der pflanzlichen Plastiden. *Helv. chim. Acta*, **48**, 1094.

Elliott, K. & Knight, J. eds. (1972). *Lipids, malnutrition and the developing brain*. Ciba Foundation Symposium, Elsevier, Amsterdam.

Ereky, K. (1927). Process for the manufacture and preservation of green fodder pulp or other plant pulp and of dry products made therefrom. *Br. Pat.* 270 629.

Erkomaishvili, S. K., Chubinidze, A. S. & Kitiashvili, D. G. (1976). Feeding value of plant juice for young cattle. Quoted from *Nutr. Abstr. Rev.*, **46**, 571.

Fafunso, M. & Bassir, O. (1976). Effects of age and season on yield of crude and extractable proteins from some edible plants. *Expl Agric.*, **12**, 249.

Fafunso, M. & Byers, M. (1977). Effect of pre-press treatments of vegetation on the quality of the extracted leaf protein. *J. Sci. Fd Agric.*, **28**, 375.

Feeny, P. P. (1969). Inhibitory effect of oak leaf tannins on the hydrolysis of proteins by trypsin. *Phytochemistry*, **8**, 2119.

Feltwell, J. S. E. & Valadon, L. R. G. (1974). Carotenoid changes in *Brassica oleracea* var. *capitata* L. with age in relation to the large white butterfly, *Pieris brassicae* L. *J. agric. Sci. Camb.*, **83**, 19.

Ferrando, R. & Spais, A. (1966). Valeur alimentaire des protéines extraites de la luzerne. *Proc. 7th Congr. Int. Nutr.*, **5**, 276.

Festenstein, G. N. (1961). Extraction of proteins from green leaves. *J. Sci. Fd Agric.*, **12**, 305.

Festenstein, G. N. (1972). Water-soluble carbohydrates in extracts from large-scale preparations of leaf protein. *J. Sci. Fd Agric.*, **23**, 1409.

Festenstein, G. N. (1976). Carbohydrates associated with leaf protein. *J. Sci. Fd Agric.*, **27**, 849.

Finot, P. A. & Mauron, J. (1972). Le blocage de la lysine par la réaction de Maillard. II. Propriétés chimiques des dérivés N-(désoxy-1-D-fructosyl-

1) et N-(désoxy-1-D-lactulosyl-1) de la lysine. *Helv. chim. Acta*, 55, 1153.

FAO (1964). *The state of food and agriculture*. FAO, Rome.

FAO (1969). *Provisional indicative world plan for agricultural development*. FAO, Rome.

FAO (1971). *Production yearbook*. FAO, Rome.

FAO (1973a). *Energy and protein requirements. Tech. Rep. Ser.*, 522. FAO, Rome.

FAO (1973b). *Food composition tables for use in East Asia*. FAO, Rome.

Foot, A. S. (1974). Lucerne juice for pigs. *Pig Fmg*, 22(9), 71.

Ford Foundation (1959). *Report on India's food crisis and steps to meet it*. Ford Foundation, New Delhi.

Foreman, F. W. (1938). Observations on the proteins of pasturage. Phosphorus and protoplasm. *J. agric. Sci. Camb.*, 28, 135.

Fourcroy, A. F. (1789). Mémoire sur l'existence de la matière albumineuse dans les végétaux. *Annls Chim.*, 3, 252.

Franzen, K. L. & Kinsella, J. E. (1976). Functional properties of succinylated and acetylated leaf protein. *J. agric. Fd Chem.*, 24, 914.

Freeman, P. A. (1977). The Simon–Barron prototype wet crop fractionation plant. In *Green crop fractionation*, ed. R. J. Wilkins, p. 167. Occasional Symposia 9, Br. Grassld Soc., Maidenhead.

Garcha, J. S., Kawatra, B. L., Wagle, D. S. & Bhatia, I. S. (1970). Studies on extraction and isolation of leaf protein of various crops grown in the Punjab. *J. Res. Punjab agric. Univ.*, 7, 211.

Garcha, J. S., Kawatra, B. L. & Wagle, D. S. (1971). Nutritional evaluation of leaf proteins and the effect of their supplementation to wheat flour by rat feeding. *J. Fd Sci. Technol.*, 8, 23.

Gastineau, C. (1974). The French work. In *Proc. 12th Tech. alfalfa conf. USDA*, p. 123. USDA, Washington.

Gastineau, C. (1976). Advertising material from France-Luzerne. France-Luzerne, 173 Avenue de Alliés, 51 Chalons S/Marne, France.

Gerloff, E. D., Lima, I. H. & Stahmann, M. A. (1965). Amino acid composition of leaf protein concentrates. *J. agric. Fd Chem.*, 13, 139.

Ghosh, J. J. (1967). Leaf protein concentrates: problems and prospects in the control of protein malnutrition. *Trans. Bose Res. Inst. (Calcutta)*, 30, 215.

Gjøen, A. U. & Njaa, L. R. (1977). Methionine sulphoxide as a source of sulphur-containing amino acids for the young rat. *Br. J. Nutr.*, 37, 93.

Glencross, R. G., Festenstein, G. N. & King, H. G. C. (1972). Separation and determination of isoflavones in the protein concentrate from red clover leaves. *J. Sci. Fd Agric.*, 23, 371.

Gonzalez, O. N., Dimaunahan, L. B. & Banyon, E. A. (1968). Extraction of protein from the leaves of some local plants. *Philipp. J. Sci.*, **97**, 17.

Goodall, C. (1936). Improvements relating to the treatment of grass and other vegetable substances. *Br. Pat.* 457 789.

Goodall, M. (1950). New high protein feed produced from sugar beet tops. *Br. Sug. Beet Rev.*, **19**, 54.

Gordon, C. H., Holdren, R. D. & Derbyshire, J. C. (1969). Field losses in harvesting wilted forage. *Agron. J.*, **61**, 924.

Gore, S. B. & Joshi, R. N. (1972). The exploitation of weeds for leaf protein. In *Symposium on tropical ecology*, eds P. M. Golley & S. B. Golley, p. 137. Athens, USA.

Gore, S. B. & Joshi, R. N. (1976a). Effect of fertiliser and frequency of cutting on the extraction of protein from *Sesbania*. *Indian J. Agron.*, **21**, 39.

Gore, S. B. & Joshi, R. N. (1976b). A note on the effect of simazine on the yields of dry matter and crude protein, and on the extractability of protein from hybrid napier grass. *Indian J. Agron.*, **21**, 491.

Gore, S. B., Mungikar, A. M. & Joshi, R. N. (1974). The yields of extracted protein from hybrid napier grass. *J. Sci. Fd Agric.*, **25**, 1149.

Green, J., Corbet, S. A., Watts, E. & Lan, O. B. (1976). Ecological studies on Indonesian lakes. Overturn and restratification of Ranu Lamongan. *J. Zool.*, **180**, 315.

Green, T. R. & Ryan, C. A. (1972). Wound-induced proteinase inhibitor in plant leaves: a possible defense mechanism against insects. *Science*, **175**, 776.

Greenhalgh, J. F. D. & Reid, G. W. (1975). Mechanical processing of wet roughage. *Proc. Nutr. Soc.*, **34**, 74A.

Guha, B. C. (1960). Leaf protein as a human food. *Lancet*, **i**, 705.

Hageman, R. H. & Waygood, E. R. (1959). Methods for the extraction of enzymes from cereal leaves with especial reference to the triosephosphate dehydrogenases. *Pl. Physiol.*, **34**, 396.

Hartley, H. (1937). Agriculture as a potential source of raw materials for industry. *J. Text. Inst.*, **28**, 151.

Hartman, G. H., Akeson, W. R. & Stahmann, M. A. (1967). Leaf protein concentrate by spray-drying. *J. agric. Fd Chem.*, **15**, 74.

Heath, S. B. (1977). The production of leaf protein concentrates from forage crops. In *Plant proteins*, ed. G. Norton, p. 171. Butterworth.

Heath, S. B. & King, M. W. (1977). The production of crops for green crop fractionation. In *Green crop fractionation*, ed. R. J. Wilkins, p. 9. Occasional Symposia 9, Br. Grassld Soc., Maidenhead.

Heidelberger, M., Kendall, F. E. & Scherp, H. W. (1936). The specific polysaccharides of types I, II and III pneumococcus. A revision of methods and data. *J. exp. Med.*, **64**, 559.

Henry, K. M. & Ford, J. E. (1965). The nutritive value of leaf protein concentrates determined in biological tests with rats and by microbiological methods. *J. Sci. Fd Agric.*, **16**, 425.

Herndon, J. H., Steinberg, D., Uhlendorf, B. W. & Fales, H. M. (1969). Refsum disease (HAP): characterisation of enzyme defect in cell culture. *J. clin. Invest.*, **48**, 1017.

Herrick, A. M. & Brown, C. L. (1967). Silage sycamore. *Agric. Sci. Rev.*, **4**, 8.

Hinde, R. & Smith, D. C. (1975). Role of photosynthesis in the nutrition of the mollusc *Elysia viridis. Biol. J. Linnean Soc.*, **7**, 161.

Holden, M. (1945). Acid-producing mechanisms in minced leaves. *Biochem. J.*, **39**, 172.

Holden, M. (1952). The fractionation and enzymic breakdown of some phosphorus compounds in leaf tissue. *Biochem. J.*, **51**, 433.

Holden, M. (1974). Chlorophyll degradation products in leaf protein preparations. *J. Sci. Fd Agric.*, **25**, 1427.

Holden, M. & Tracey, M. V. (1948). The effect of fertilizers on the levels of nitrogen, phosphorus, protease, and pectase in healthy tobacco leaves. *Biochem. J.*, **43**, 147.

Holden, M. & Tracey, M. V. (1950). A study of enzymes that can break down tobacco-leaf components. 4. Mammalian pancreatic and salivary enzymes. *Biochem. J.*, **47**, 421.

Holl, W. & Hampp, R. (1975). Lead in plants. *Residue Rev.*, **54**, 79.

Hollo, J. & Koch, L. (1971). Commercial production in Hungary. In *Leaf protein: its agronomy, preparation, quality and use*, ed. N. W. Pirie, p. 63. Blackwell, Oxford.

Horigome, T. & Kandatsu, M. (1964). Studies on the nutritive value of grass proteins: XII. Digestibilities of isolated grass proteins. *J. agric. Chem. Soc. Japan*, **38**, 121.

Horn, M. J., Lichtenstein, H. & Womack, M. (1968). Availability of amino acids: a methionine–fructose compound and its availability to microorganisms and rats. *J. agric. Fd Chem.*, **16**, 741.

Houseman, R. A. & Connell, J. (1976). The utilization of the products of green-crop fractionation by pigs and ruminants. *Proc. Nutr. Soc.*, **35**, 213.

Hove, E. L., Lohrey, E., Urs, M. K. & Allison, R. M. (1974). The effect of lucerne-protein concentrate in the diet on growth, reproduction and body composition of rats. *Br. J. Nutr.*, **31**, 147.

Hove, E. L. & Bailey, R. W. (1975). Towards a leaf protein concentrate industry in New Zealand. *N.Z. J. exp. Agric.*, **3**, 193.

Huang, K. H., Tao, M. C., Boulet, M., Riel, R. R., Julien, J. P. & Brisson, G. J. (1971). A process for the preparation of leaf protein concentrates based on the treatment of leaf juices with polar solvents. *Can. Inst. Fd Technol.*, **4**(3), 85.

Hudson, B. J. F. & Karis, I. G. (1973). Aspects of vegetable structural lipids: I. The lipids of leaf protein concentrate. *J. Sci. Fd Agric.*, **24**, 1541.

Hudson, B. J. F. & Karis, I. G. (1974). Aspects of vegetable structural lipids: II. The effect of crop maturity on leaf lipids. *J. Sci. Fd Agric.*, **25**, 1491.

Hudson, B. J. F. & Warwick, M. J. (1977). Lipid stabilisation in leaf protein concentrates from ryegrass. *J. Sci. Fd Agric.*, **28**, 259.

Hudson, J. P. (1975). Weeds as crops. In *Proc. 12th Br. weed control conf.*, p. 333. Br. Crop Protection Coun., London.

Hudson, J. P. (1976). Food crops for the future. *J. R. Soc. Arts*, **124**, 572.

Hume, E. M. (1921). Comparison of the growth-promoting properties for guinea-pigs of certain diets, consisting of natural foodstuffs. *Biochem. J.*, **15**, 30.

Hume, E. M. & Krebs, H. A. (1949). Vitamin A requirement of human adults. *Med. Res. Coun. Spec. Rep.*, **264**.

Hussain, A., Ullah, M. & Ahmad, B. (1968). Studies on the potentials of leaf proteins for the preparation of concentrates from various leaf-wastes in West Pakistan. *Pakist. J. agric. Res.*, **6**, 110.

Ivins, J. D. (1973). Increasing productivity: crop physiology and nutrition. *Phil. Trans. R. Soc. Lond.*, B, **267**, 81.

Jadhav, B. & Joshi, R. N. (1977). Extractability of leaf protein from the green tops of groundnut (*Arachis hypogaea* L.). *J. Fd Sci. Technol.*, **14**, 179.

Jennings, A. C. & Watt, W. B. (1967). Fractionation of plant material. I. Extraction of proteins and nucleic acids from plant tissues and isolation of protein fractions containing hydroxyproline from broad bean (*Vicia faba* L.) leaves. *J. Sci. Fd Agric.*, **18**, 527.

Jennings, A. C., Pusztai, A., Synge, R. L. M. & Watt, W. B. (1968). Fractionation of plant material. III. Two schemes for chemical fractionation of fresh leaves, having special applicability for isolation of the bulk protein. *J. Sci. Fd Agric.*, **19**, 203.

Jones, R. J., Blunt, C. G. & Holmes, J. G. H. (1976). Enlarged thyroid glands in cattle grazing Leucaena pastures. *Trop. Grassl.*, **10**, 113.

Jones, W. T. & Mangan, J. L. (1976). Large-scale isolation of fraction 1 leaf protein (18 S) from lucerne (*Medicago sativa* L.). *J. agric. Sci. Camb.*, **86**, 495.

Jönsson, A. G. (1962). Studies in the utilization of some agricultural wastes and by-products by various microbial processes. *K. LantbrHögsk. Annlr.*, **28**, 235.

Joshi, R. N. (1971). The yields of leaf protein that can be extracted from crops of Aurangabad. In *Leaf protein: its agronomy, preparation, quality and use*, ed. N. W. Pirie, p. 19. Blackwell, Oxford.

Kamalanathan, G. & Devadas, R. P. (1971). Acceptability of food preparations containing leaf protein concentrates. In *Leaf protein: its agronomy, preparation, quality and use*, ed. N. W. Pirie, p. 145. Blackwell, Oxford.

Kamalanathan, G., Karuppiah, P. & Devadas, R. P. (1975). Supplementary value of leaf protein and groundnut meal in the diets of preschool children. *Indian J. Nutr. Dietet.*, **12**, 203.

Kanev, S., Boncheva, I., Georgieva, L. & Iovchev, N. (1976). Tests of a protein concentrate from lucerne for fattening pigs and poultry. Quoted from *Nutr. Abstr. Rev.*, **46**, 282.

Kannangara, C. G. & Stumpf, P. K. (1972). Fat metabolism in higher plants. I. The biosynthesis of polyunsaturated fatty acids by isolated spinach chloroplasts. *Arch. Biochim. biophys.*, **148**, 414.

Kawatra, B. L., Garcha, J. S. & Wagle, D. S. (1974). Effect of supplementation of leaf protein extracted from berseem (*Trifolium alexandrinum*) to wheat flour diet. *J. Fd Sci. Technol.*, **11**, 241.

Kempton, T. J., Nolan, J. V. & Leng, R. A. (1977). Principles for the use of non-protein nitrogen and by-pass protein in diets of ruminants. *Wld Anim. Rev.*, **22**, 2.

Kiesel, A., Belozersky, A., Agatov, P., Biwshich, N. & Pawlowa, M. (1934). Vergleichende Untersuchungen über Organeiweiss von Pflanzen. *Z. physiol. Chem.*, **226**, 73.

Knuckles, B. E., de Fremery, D., Bickoff, E. M. & Kohler, G. O. (1975). Soluble protein from alfalfa juice by membrane filtration. *J. agric. Fd Chem.*, **23**, 209.

Knuckles, B. E., de Fremery, D. & Kohler, G. O. (1976). Coumestrol content of fractions obtained during wet processing of alfalfa. *J. agric. Fd Chem.*, **24**, 1177.

Koch, L. (1974). Producing protein concentrates from green plants. *Växtodling* (*Plant Husbandry*), **28**, 129.

Koegel, R. G. & Bruhn, H. D. (1977). An analysis of the requirements for expression of plant juice. In *Green crop fractionation*, ed. R. J. Wilkins, p. 23. Occasional Symposia 9, Br. Grassld Soc., Maidenhead.

Kotake, Y. & Knoop, F. (1911). Ueber einen krystallisierten Eiweiss-körper aus dem Milchsäfte der *Antiaris toxicaria*. *Z. physiol. Chem.*, **75**, 488.

Laird, W. M., Mbadiwe, E. I. & Synge, R. L. M. (1976). A simplified procedure for fractionating plant materials. *J. Sci. Fd Agric.*, **27**, 127.

Lala, V. R. & Reddy, V. (1970). Absorption of β carotene from green leafy vegetables in undernourished children. *Am. J. clin. Med.*, **23**, 110.

Lawes, J. B. (1864). On the chemistry of the feeding of animals for the production of meat and manure. *Farmers Gazette*, Dublin.

Lawes, J. B. (1885). Sugar as a food for stock. *Jl R. agric. Soc.*, **21**, 81.

Lea, C. H. & Parr, L. J. (1961). Some observations on the oxidative deterioration of the lipids of crude leaf protein. *J. Sci. Fd Agric.*, **12**, 785.

Levin, D. A. (1971). Plant phenolics: an ecological perspective. *Am. Nat.*, **105**, 157.

Levin, D. A. (1976). The chemical defenses of plants to pathogens and herbivores. *A. Rev. Ecol. Syst.*, **7**, 121.

Lexander, K., Carlsson, R., Schalen, V., Simonsson, A. & Lundborg, T. (1970). Quantities and qualities of leaf protein concentrates from wild species and crop species grown under controlled conditions. *Ann. appl. Biol.*, **66**, 193.

Lima, I. H., Richardson, T. & Stahmann, M. A. (1965). Fatty acids in some leaf protein concentrates. *J. agric. Fd Chem.*, **13**, 143.

Little, E. C. S. (1968). *Handbook of utilization of aquatic plants*. FAO, Rome.

Lohrey, E., Tapper, B. & Hove, E. L. (1974). Photosensitization of albino rats fed on lucerne-protein concentrates. *Br. J. Nutr.*, **31**, 159.

Lugg, J. W. H. (1932). The application of phospho-18-tungstic acid (Folin's reagent) to the colorimetric determination of cysteine, cystine and related substances. II. The determination of sulphydryl compounds and disulphides already existing in solution. *Biochem. J.*, **26**, 2160.

Lugg, J. W. H. (1939). The representativeness of extracted samples and the efficiency of extraction of protein from the fresh leaves of plants; and some partial analyses of the whole proteins of leaves. *Biochem. J.*, **33**, 110.

Magoon, M. L. (1972). March towards self-sufficiency in animal food. *Indian Fmg*, **22**, 25.

Maguire, M. F. & Brookes, I. M. (1972). Grass juice as a liquid feed. *Res. Rep. An Foras Taluntais. Anim. Prod.*, 38.

Maguire, M. F. & Brookes, I. M. (1973). The effects of juice extraction on the composition and yield of grass crops for dehydration. In *Proc. 1st int. congr. green crop drying*, p. 346. Ass. Green Crop Driers, Oxford.

Mahadeviah, S. & Singh, N. (1968). Leaf protein from the green tops of *Cichorium intybus* L. (chicory). *Indian J. exp. Biol.*, **6**, 193.

Marchaim, U., Birk, Y., Dovrat, A. & Berman, T. (1972). Lucerne saponins as inhibitors of cotton seed germination: their effect on diffusion of oxygen through seed coats. *J. exp. Bot.*, **23**, 302.

Martin, C. (1975). Leaf protein child feeding trial. *League Int. Fd Educ. Newsl.*, March.

Matai, S. (1976). Protein from water weeds. In *Aquatic weeds in South East Asia*, ed. C. K. Varshney & J. Rzoska, p. 369. Junk, The Hague.

Matai, S. & Bagchi, D. K. (1974). Some promising legumes for leaf protein extraction. *Sci. Cult.*, **40**, 34.

Matai, S., Bagchi, D. K. & Roy Chowdhury, S. (1971). Leaf protein from some plants in West Bengal. *Sci. Engng.*, **24**, 102.

Matai, S., Bagchi, D. K. & Chanda, S. (1973). Optimal seed rate and fertilizer dose for maximum yield of extracted protein from the leaves of mustard (*Brassica nigra* Koch) and turnip (*Brassica rapa* L.). *Indian J. agric. Sci.*, **43**, 165.

Matai, S., Bagchi, D. K. & Chanda, S. (1976). Effects of seed rate, nitrogen level and leaf age on the yield of extracted protein from five different crops in West Bengal. *J. Sci. Fd Agric.*, **27**, 736.

Mathismoen, P. (1974). Stord twin screw presses: design and application on alfalfa. In *Proc. 12th tech. alfalfa conf.*, p. 135. USDA, Washington.

Mauron, J. (1970a). Le comportement chimique des protéines lors de la préparation des aliments et ses incidences biologiques. *J. int. Vitaminol.*, **40**, 209.

Mauron, J. (1970b). Nutritional evaluation of proteins by enzymatic methods. In *Evaluation of novel protein products*, ed. A. E. Bender, R. Kihlberg, B. Löfquist & L. Munk, p. 211. Pergamon.

McClure, J. W. (1970). Secondary constituents of aquatic angiosperms. In *Phytochemical phylogeny*, ed. J. B. Harborne, p. 233. Academic Press, London.

McLeod, M. N. (1974). Plant tannins – their role in forage quality. *Nutr. Abstr. Rev.*, **44**, 804.

158 *References*

Mead, J. F. & Alfin-Slater, R. B. (1966). Toxic substances present in food fats. *Natn. Acad. Sci./Natn. Res. Coun. Publ.*, **1354**, p. 238.

Milic, B. L. (1972). Lucerne tannins. I. Content and composition during growth. *J. Sci. Fd Agric.*, **23**, 1151.

Milic, B. L. & Stojanovic, S. (1972). Lucerne tannins. III. Metabolic fate of lucerne tannins in mice. *J. Sci. Fd Agric.*, **23**, 1163.

Miller, D. S. (1965). Some nutritional problems in the utilization of non-conventional proteins for human feeding. *Recent Adv. Fd Sci.*, **3**, 125.

Miller, D. S. & Samuel, P. D. (1970). Effects of addition of sulphur compounds to the diet on utilisation of protein in young growing rats. *J. Sci. Fd Agric.*, **21**, 616.

Miller, R. E., Edwards, R. H., Lazar, M. E., Bickoff, E. M. & Kohler, G. O. (1972). PRO-XAN process: air drying alfalfa leaf protein concentrate. *J. agric. Fd Chem.*, **20**, 1151.

Miller, R. E., de Fremery, D., Bickoff, E. M. & Kohler, G. O. (1975). Soluble protein concentrate from alfalfa by low-temperature acid precipitation. *J. agric. Fd Chem.*, **23**, 1177.

Mitchell, D. S., ed. (1974). *Aquatic vegetation and its use and control.* UNESCO, Paris.

Mitscher, L. A. (1975). Antimicrobial agents from higher plants. *Recent Adv. Phytochem.*, **9**, 243.

Mokady, S. & Zimmermann, C. (1966). The effect of different liquid extractants used with impulse-rendered lucerne paste on the nutritional and calorigenic properties of the proteins. *Proc. 7th Int. Congr. Nutr.*, **5**, 279.

Monsod, G. G. (1976). *The versatility and economics of water hyacinth.* Phillip. Coun. agric. Resources Res., Fisheries Forum, Manila.

Moore, P. D. (1976). Chemical antagonism in plant communities. *Nature*, **259**, 447.

Morris, T. R. (1977). Leaf protein concentrate for non-ruminant animals. In *Green crop fractionation*, ed. R. J. Wilkins, p. 67. Occasional Symposia 9, Br. Grassld Soc., Maidenhead.

Morrison, J. E. & Pirie, N. W. (1960). The presentation of leaf protein on the table. *Nutrition*, **14**, 7.

Morrison, J. E. & Pirie, N. W. (1961). The large scale production of protein from leaf extracts. *J. Sci. Fd Agric.*, **12**, 1.

Mungikar, A. M. & Joshi, R. N. (1976). Studies on the ensilage of the residues left after extraction of leaf protein from lucerne and hybrid napier grass. *Indian J. Nutr. Dietet.*, **13**, 39.

Mungikar, A. M., Tekale, N. S. & Joshi, R. N. (1976a). The yields of leaf protein and fibre that can be obtained from fractionation of berseem (*Trifolium alexandrinum* L.). *Indian J. Nutr. Dietet.*, **13**, 114.

Mungikar, A. M., Batra, U. R., Tekale, N. S. & Joshi, R. N. (1976b). Effects of nitrogen fertilisation on the yields of extracted protein from some crops. *Expl Agric.*, **12**, 353.

Munshi, S. K., Wagle, D. S. & Thapar, V. K. (1974). Nutritional evaluation of leaf protein isolates treated under different drying temperatures. *Labdev J. Sci. Technol. ind.*, **12B**, 35.

Myer, R. O., Cheeke, P. R. & Kennick, W. H. (1975). Utilization of alfalfa protein concentrate by swine. *J. Anim. Sci.*, **40**, 885.

Nass, M. M. K. (1969). Uptake of isolated chloroplasts by mammalian cells. *Science*, **165**, 1128.

National Academy of Sciences (1976). *Making aquatic weeds useful: some perspectives for developing countries*. NAS, Washington.

National Economic Development Office (1974). *UK farming and the Common Market: grass and grass products*. NEDO, London.

Naumenko, V., Tarasenko, A. & Kinsburgskii, Z. (1975). Juice from lucerne in feeds for young pigs. Quoted from *Nutr. Abstr. Rev.*, **45**, 580·

Naumenko, V., Morozova, A. & Kinsburgskii, Z. (1977). Lucerne juice in diets for pigs. Quoted from *Nutr. Abstr. Rev.*, **47**, 341.

Nazir, M. & Shah, F. H. (1966). Extractability of proteins from various leaves. *Pakist. J. scient. ind. Res.*, **9**, 235.

Nedorizescu, M. (1972). Production of some fodder meals made from different forest products. Quoted from *Nutr. Abstr. Rev.*, 1974, **44**, 234.

Njaa, L. R. (1962). Some problems related to detection of methionine sulfoxide in protein hydrolysates. *Acta chem. scand.*, **16**, 1359.

Nowakowski, T. Z. & Byers, M. (1972). Effects of nitrogen and potassium fertilisers on contents of carbohydrates and free amino acids in Italian ryegrass. II. Changes in the composition of the non-protein nitrogen fraction and the distribution of individual amino acids. *J. Sci. Fd Agric.*, **23**, 1313.

Nutrition Society (1973). Fibre in human nutrition. *Proc. Nutr. Soc.*, **32**, 123.

Oelshlegel, F. J., Schroeder, J. R. & Stahmann, M. A. (1969). Potential for protein concentrates from alfalfa and waste green plant material. *J. agric. Fd Chem.*, **17**, 791.

Oke, O. L. (1966). The introduction of leaf protein into the Nigerian diet. *Nutrition*, **20**, 18.

Olatunbosun, D. A. (1976). Leaf protein for human use in Africa. *Indian J. Nutr. Dietet.*, **13**, 168.

Olatunbosun, D. A., Adadevoh, B. K. & Oke, O. L. (1972). Leaf protein: a new protein source for the management of protein calorie malnutrition in Nigeria. *Niger. med. J.*, **2**, 195.

Omole, T. A., Oke, O. L. & Mfon, B. P. (1976). Carrot leaf protein: preliminary trials with whole leaf using rabbits. *Nutr. Rep. Int.*, **14**, 173.

Ornes, W. H. & Sutton, D. I. (1975). Removal of phosphorus from static sewage effluent by water hyacinth. *Hyacinth Contr. J.*, **13**, 56.

Osborne, T. B. (1924). *The vegetable proteins*, 2nd ed. Longmans, London.

Osborne, T. B. & Wakeman, A. J. (1920). The proteins of green leaves. I. Spinach leaves. *J. biol. Chem.*, **42**, 1.

Osborne, T. B., Wakeman, A. J. & Leavenworth, C. S. (1921). The proteins of the alfalfa plant. *J. biol. Chem.*, **49**, 63.

Oyakawa, N., Orlandi, W. & Valente, E. O. L. (1968). The use of *Eichhornia crassipes* in the production of yeast, feeds and forages. Quoted from Little (1968).

Paredes-Lopez, O. & Camagro, E. (1973). The use of alfalfa residual juice for production of single-cell protein. *Experientia*, **29**, 1233.

Patriquin, D. G. & Knowles, R. (1972). Nitrogen fixation in the rhizosphere of marine angiosperms. *Mar. Biol.*, **16**, 49.

Pereira, S. M. & Begum, A. (1968). Studies in the prevention of vitamin A deficiency. *Indian J. med. Res.*, **56**, 362.

Peterson, D. W. (1950). Some properties of a factor in alfalfa meal causing depression of growth in chicks. *J. biol. Chem.*, **183**, 647.

Pienazek, D., Rakowska, M. & Kunachowicz, H. (1975). The participation of methionine and cysteine in the formation of bonds resistant to the action of proteolytic enzymes in heated casein. *Br. J. Nutr.*, **34**, 163.

Pierpoint, W. S. (1959). Mitochondrial preparations from the leaves of tobacco (*Nicotiana tabacum*). *Biochem. J.*, **71**, 518.

Pierpoint, W. S. (1969). *o*-Quinones formed in plant extracts. Their reactions with amino acids and peptides. *Biochem. J.*, **112**, 609.

Pierpoint, W. S. (1971). Formation and behaviour of *o*-quinones in some processes of agricultural importance. *A. Rep. Rothamsted exp. Stn*, 1970, 199.

Pieterse, A. H. (1974). The water hyacinth. *Trop. Abstr.*, **29**, 77.

Pirie, A. (1975). High carotene Indonesian leaf vegetables. *Xerophthalmia Club Bull.*, **8**, 2. (Available from A. Pirie, Nuffield Laboratory of Ophthalmology, Oxford.)

Pirie, N. W. (1942*a*). Direct use of leaf protein in human nutrition. *Chemy Ind.*, **61**, 45.

Pirie, N. W. (1942*b*). Green leaves as a source of proteins and other nutrients. *Nature*, **149**, 251.

Pirie, N. W. (1950). The isolation from normal tobacco leaves of nucleoproteins with some similarity to plant viruses. *Biochem. J.*, **47**, 614.

Pirie, N. W. (1951). The circumvention of waste. In *Four thousand million mouths*, ed. F. LeGros Clark & N. W. Pirie, p. 180. Oxford Univ. Press, Oxford.

Pirie, N. W. (1953). Large-scale production of edible protein from fresh leaves. *A. Rep. Rothamsted exp. Stn*, 1952, 173.

Pirie, N. W. (1955). Proteins. In *Modern methods of plant analysis*, ed. K. Paech & M. V. Tracey, vol. 4, p. 23. Springer, Heidelberg.

Pirie, N. W. (1957). Biochemical Engineering. *Research*, **10**, 29.

Pirie, N. W. (1958). Large-scale production of leaf protein. *A. Rep. Rothamsted exp. Stn*, 1957, 102.

Pirie, N. W. (1959*a*). Leaf proteins. *A. Rev. Pl. Physiol.*, **10**, 33.

Pirie, N. W. (1959*b*). The large-scale separation of fluids from fibrous pulps. *J. biochem. microbiol. Technol. Engng*, **1**, 13.

Pirie, N. W. (1959*c*). Large-scale production of leaf protein. *A. Rep. Rothamsted exp. Stn*, 1958, 93.

Pirie, N. W. (1961). The disintegration of soft tissues in the absence of air. *J. agric. Engng Res.*, **6**, 142.

Pirie, N. W. (1962). Progress in biochemical engineering broadens our choice of crop plants. *Econ. Bot.*, **15**, 302.

Pirie, N. W. (1963). The selection and use of leafy crops as a source of protein for man. *Proc. 5th int. Congr. Biochem.*, **8**, 53.

Pirie, N. W. (1964*a*). The size of small organisms. *Proc. R. Soc.*, B, **160**, 149.

Pirie, N. W. (1964*b*). Large-scale production of leaf protein. *A. Rep. Rothamsted exp. Stn*, 1963, 96.

Pirie, N. W. (1964*c*). Freeze-drying, or drying by sublimation. In *Instrumental methods of experimental biology*, ed. D. W. Newman, p. 189. Macmillan, New York.

Pirie, N. W. (1966*a*). Improvements to machinery. *A. Rep. Rothamsted exp. Stn*, 1965, 106.

Pirie, N. W. (1966*b*). Fodder fractionation: an aspect of conservation. *Fertil. Feed. Stuffs J.*, **63**, 119.

Pirie, N. W. (1970*a*). Large-scale protein preparations. *A. Rep. Rothamsted exp. Stn*, 1969, 132.

Pirie, N. W. (1970b). Weeds are not all bad. *Ceres*, 3(4), 31.

Pirie, N. W., ed. (1971a). *Leaf protein: its agronomy, preparation, quality and use*. Blackwell, Oxford.

Pirie, N. W. (1971b). Some obstacles to innovation. *Pugwash Newsl.*, 9, 16.

Pirie, N. W. (1972). The direction of beneficial nutritional change. *Ecol. Fd Nutr.*, 1, 279.

Pirie, N. W. (1974). Fixation of nucleic acid by leaf fibre and calcium phosphate. *Proc. R. Soc.*, B, 185, 343.

Pirie, N. W. (1975a). The potentialities of leafy vegetables and forages as food protein sources. *Baroda J. Nutr.*, 2, 43.

Pirie, N. W. (1975b). Some obstacles to eliminating famine. *Proc. Nutr. Soc.*, 34, 181.

Pirie, N. W. (1976a). *Food resources: conventional and novel*. Penguin Books, Harmondsworth.

Pirie, N. W. (1976b). Food protein sources. *Phil. Trans. R. Soc.*, B, 274, 489.

Pirie, N. W. (1976c). Restoring esteem for leafy vegetables. *Appropriate Technol.*, 3(3), 24.

Pirie, N. W. (1977a). The extended use of fractionation processes. *Phil. Trans. R. Soc.*, B, 281, 139.

Pirie, N. W. (1977b). A simple unit for extracting leaf protein in bulk. *Expl Agric.*, 13, 113.

Pirie, N. W., Fairclough, D. & Shardlow, A. W. (1958). Improvements in apparatus for the expulsion of fluid from fibrous materials. *Br. Pat.* 800 778.

Pleshkov, B. P. & Fowden, L. (1959). Amino-acid composition of the proteins of barley leaves in relation to the mineral nutrition and age of plants. *Nature*, 183, 1445.

Powling, W. T. (1953). Protein extraction from green crops. *Wld Crops*, 5, 63.

Prasad, V. L., Dev, D. V., Patil, R. E., Joshi, A. L. & Rangnekar, D. V. (1977). A note on feeding of lucerne extract to preruminant calves as part of milk replacer. *Indian J. Dairy Sci.*, 30, 154.

Protein Advisory Group (1970). Statement on leaf protein concentrate. *PAG Bull.*, 10, 3.

Proust, J. L. (1803). An essay on the fecula of green plants. *Phil. Mag.*, 16, 122 and 17, 22.

Ramana, K. V. R. & Singh, N. (1971). Nutritional efficiency of lucerne leaf protein as a source of β carotene in rat diets. *Indian J. exp. Biol.*, 9, 378.

Randolph, J. W., Rivera-Brenes, L., Winfree, J. P. & Green, V. E. (1958). Mechanical dewatering as a potential means for improving the supply of quality animal feeds in the tropics and sub-tropics. *Proc. Soil Crop Sci. Soc. Florida*, **18**, 97.

Rao, C. N. & Rao, B. S. (1970). Absorption of dietary carotenes in human subjects. *Am. J. clin. Nutr.*, **23**, 105.

Raymond, W. F. (1977). Farm wastes. *Biologist*, **24**, 80.

Raymond, W. F. & Harris, C. E. (1957). The value of the fibrous residue from leaf-protein extraction as a feeding stuff for ruminants. *J. Br. Grassld Soc.*, **12**, 166.

Ream, H. W., Smith, D. & Walgenbach, R. P. (1977). Effects of deproteinized alfalfa juice applied to alfalfa–bromegrass, bromegrass and corn. *Agron. J.*, **69**, 685.

Richardson, M. (1977). The proteinase inhibitors of plants and microorganisms. *Phytochemistry*, **16**, 159.

Ries, S. K., Wert, V., Sweely, C. C. & Leavitt, R. A. (1977). Triacontanol: a new naturally occurring plant growth regulator. *Science*, **195**, 1339.

Roberts, E. A. H. (1959). The interaction of flavonol orthoquinones with cysteine and glutathione. *Chemy Ind.*, 995.

Robson, T. O. (1973). *The control of aquatic weeds. Min. Agric. Fish. Fd Bull.*, **194**.

Rodriguez, E., Towers, G. H. N. & Mitchell. J. C. (1977). Allergic contact dermatitis and sesquiterpene lactones. *Compositae Newsl.*, **4**, 4.

Roja, F. C. & Smith, R. A. (1977). The antibacterial screening of some common ornamental plants. *Econ. Bot.*, **31**, 28.

Rothschild, M., Valadon, G. & Mummery, R. (1977). Carotenoids of the pupae of the Large White butterfly (*Pieris brassicae*) and the Small White butterfly (*Pieris rapae*). *J. Zool.*, **181**, 323.

Rouelle, H. M. (1773). Observations sur les fécules ou parties vertes des plantes, & sur la matière glutineuse ou végéto-animale. *J. Méd. Chir. Pharm.*, **40**, 59.

Sale, P. J. M. (1973). Productivity of vegetable crops in a region of high solar input. I. Growth and development of the potato. *Aust. J. agric. Res.*, **24**, 733; 751.

Saunders, R. M., Connor, M. A., Booth, A. N., Bickoff, E. M. & Kohler, G. O. (1973). Measurement of digestibility of alfalfa protein concentrates by *in vivo* and *in vitro* methods. *J. Nutr.*, **103**, 530.

Savangikar, V. A. & Joshi, R. N. (1976). Influence of irrigation and fertiliser on the yields of extracted protein from lucerne. *Forage Res.*, **2**, 125.

Savangikar, V. A. & Joshi, R. N. (1978). Edible protein from *Parthenium hysterophorus* L. *Expl Agric.*, **14**, 93.

Savel'ev, D. N. O. (1970). Changes in carotene contents of fodder plants. Quoted from *Nutr. Abstr. Rev.*, **40**, 42.

Schnabel, C. F. (1938). Vitaminic product from grass juice. *US Pat.* 2 133 362.

Schubert, K. R. & Evans, H. J. (1976). Hydrogen evolution: a major factor affecting the efficiency of nitrogen fixation in nodulated symbionts. *Proc. natn. Acad. Sci. U.S.A.*, **73**, 1207.

Schwarz, K. (1977). Silicon, fibre, and atherosclerosis. *Lancet*, **i**, 454.

Scrimshaw, N. S. (1976). Strengths and weaknesses of the committee approach. *New Engl. J. Med.*, **294**, 136; 198.

Sentheshanmuganathan, S. & Durand, S. (1969). Isolation and composition of proteins from leaves of plants grown in Ceylon. *J. Sci. Fd Agric.*, **20**, 603.

Shah, F. H. (1968). Changes in leaf protein lipids *in vitro*. *J. Sci. Fd Agric.*, **19**, 199.

Shah, F. H. (1971). Effect of heat on the extractability of lipid from leaf protein meal. *Pakist. J. scient. ind. Res.*, **14**, 492.

Shah, F. H., Riaz-ud-Din & Salam, A. (1967). Effect of heat on the digestibility of leaf proteins. I. Toxicity of the lipids and their oxidation products. *Pakist. J. scient. ind. Res.*, **10**, 39.

Shah, F. H., Zia-ur-Rehman & Mahmud, B. A. (1976). Effect of extraction techniques on the extraction of protein from *Trifolium alexandrinum*. *Pakist. J. scient. ind. Res.*, **19**, 39.

Shumway, L. K., Rancour, J. M. & Ryan, C. A. (1970). Vacuolar protein bodies in tomato leaf cells and their relationship to storage of chymotrypsin inhibitor 1 protein. *Planta*, **93**, 1.

Shurpalekar, K. S., Singh, N. & Sundaravalli, O. E. (1969). Nutritive value of leaf protein from lucerne (*Medicago sativa*): growth responses in rats at different protein levels and to supplementation with lysine and/or methionine. *Indian J. exp. Biol.*, **7**, 279.

Singh, N. (1960). Differences in the nature of nitrogen precipitated by various methods from wheat leaf extracts. *Biochim. biophys. Acta*, **45**, 422.

Singh, N. (1962). Proteolytic activity of leaf extracts. *J. Sci. Fd Agric.*, **13**, 325.

Singh, N. (1964). Leaf protein extraction from some plants of northern India. *J. Fd Sci. Technol.*, **1**, 37.

Siren, G., Blombäck, B. & Alden, T. (1970). *Proteins in forest tree leaves. R. Coll. For. (Sweden) Res. Note*, **28**.

Siren, G. (1973). Protein ur skogsträd. *Svensk Naturv.*, **41**.

Skinner, F. A. (1955). Antibiotics. In *Modern methods of plant analysis*, ed. K. Paech & M. V. Tracey, vol. 3, p. 626. Springer, Heidelberg.

Slade, R. E. (1937). Grass and the national food supply. In *Br. Ass. a. Rep.*, p. 457. Br. Ass. Adv. Sci.

Slade, R. E., Birkinshaw, J. H. & ICI (1939). Improvements in or related to the utilization of grass and other green crops. *Br. Pat.* 511, 525.

Slade, R. E., Branscombe, D. J. & McGowan, J. C. (1945). Protein extraction. *Chemy Ind.*, **23**, 194.

Slansky, F. & Feeny, P. (1977). Stabilization of the rate of nitrogen accumulation by larvae of the cabbage butterfly on wild and cultivated food-plants. *Ecology Monographs*, **47**, 209.

Slesser, M., Lewis, C. & Edwardson, W. (1977). Energy systems analysis for food policy. *Fd Policy*, **2**, 123.

Smith, D. A. & Woodruff, M. F. A. (1951). *Deficiency diseases in Japanese prison camps. Med. Res. Coun. Spec. Rep.*, **274**.

Smith, R. C. (1972). Acetylation of methionine sulfoxide and methionine sulfone by the rat. *Biochim. biophys. Acta*, **261**, 304.

Spencer, R. R., Mottola, A. C., Bickoff, E. M., Clark, J. P. & Kohler, G. O. (1971). The PRO-XAN process: the design and evaluation of a pilot plant system for the coagulation and separation of the leaf protein from alfalfa juice. *J. agric. Fd Chem.*, **19**, 504.

Squibb, R. L., Mendez, J., Guzman, M. A. & Scrimshaw, N. S. (1954). Ramie; a high protein forage crop for tropical areas. *J. Br. Grassld Soc.*, **9**, 313.

Staron, T. (1975). Une méthode d'obtention des protéines à partir des plantes vertes. *C.R. hebd. Séanc. Acad. Agric. Fr.*, **61**, 446.

Subba Rao, M. S., Singh, N. & Prasanappa, G. (1967). Preservation of wet leaf protein concentrates. *J. Sci. Fd Agric.*, **18**, 295.

Subba Rau, B. H., Mahadeviah, S. & Singh, N. (1969). Nutritional studies on whole-extract coagulated leaf protein and fractionated chloroplastic and cytoplasmic proteins from lucerne (*Medicago sativa*). *J. Sci. Fd Agric.*, **20**, 355.

Subba Rau, B. H. & Singh, N. (1970). Studies on nutritive value of leaf protein from lucerne (*Medicago sativa*): II. Effect of processing conditions. *Indian J. exp. Biol.*, **8**, 34.

Subba Rau, B. H. & Singh, N. (1971). Studies on nutritive value of leaf protein from lucerne (*Medicago sativa*): III. Supplementation of rat diets based on wheat. *J. Sci. Fd Agric.*, **22**, 569.

Subba Rau, B. H., Ramana, K. V. R. & Singh, N. (1972). Studies on nutritive value of leaf proteins and some factors affecting their quality. *J. Sci. Fd Agric.*, **23**, 233.

Sullivan, J. T. (1943). Protein concentrates from grasses. *Science*, **98**, 363.

Sullivan, J. T. (1944). High-protein concentrate can be obtained from grass. *Fd Inds*, 186.

Sur, B. K. (1961). Nutritive value of lucerne-leaf proteins. Biological value of lucerne proteins and their supplementary relations to rice proteins measured by balance and rat growth methods. *Br. J. Nutr.*, **15**, 419.

Synge, R. L. M. (1975). Interactions of polyphenols with proteins in plants and plant products. *Qualitas Pl. Mater. veg.*, **24**, 337.

Synge, R. L. M. (1976). Damage to nutritional value of plant proteins by chemical reactions during storage and processing. *Qualitas Pl. Mater. veg.*, **26**, 9.

Tapper, B. A., Lohrey, E., Hove, E. L. & Allison, R. M. (1975). Photosensitivity from chlorophyll-derived pigments. *J. Sci. Fd Agric.*, **26**, 277.

Taylor, D. L. (1968). Chloroplasts as symbiotic organelles in the digestive gland of *Elysia viridis* (Gastropoda: Opisthobranchia). *J. mar. biol. Ass.*, **48**, 1.

Taylor, K. G., Bates, R. P. & Robbins, R. C. (1971). Extraction of protein from water hyacinth. *Hyacinth Contr. J.*, **9**, 20.

Tekale, N. S. & Joshi, R. N. (1976). Extractable protein from by-product vegetation of some cole and root crops. *Ann. appl. Biol.*, **82**, 155.

Thakur, M. L., Somaroo, B. H. & Grant, W. F. (1974). The phenolic constituents from leaves of *Manihot esculenta*. *Can. J. Bot.*, **52**, 2381.

Thresh, J. M. (1956). Some effects of tannic acid and of leaf extracts which contain tannins on the infectivity of tobacco mosaic and tobacco necrosis viruses. *Ann. appl. Biol.*, **44**, 608.

Thung, T. H. & van der Want, J. P. H. (1951). Viruses and tannins. *Tijdschr. PlZiekt.*, **57**, 72.

Toosy, R. Z. & Shah, F. H. (1974). Leaf protein concentrate in human diet. *Pakist. J. scient. ind. Res.*, **17**, 40.

Tracey, M. V. (1948). Leaf protease of tobacco and other plants. *Biochem. J.*, **42**, 281.

Tragardh, C. (1974). Production of leaf protein concentrate for human consumption by isopropanol treatment. A comparison between untreated raw juice and raw juice concentrated by evaporation and ultrafiltration. *Lebensm. Wiss. Technol.*, **7**, 199.

Trench, R. K., Boyle, J. E. & Smith, D. C. (1973). The association between chloroplasts of *Codium fragile* and the mollusc *Elysia viridis*. I. Characteristics of isolated *Codium* chloroplasts. *Proc. R. Soc., B*, **184**, 51.

Trigg, T. E. & Topps, J. H. (1971). The effects of additional methionine on the quality of leaf protein concentrates differing in nutritive value. *Proc. Nutr. Soc.*, **31**, 45A.

Tsuchihashi, M. (1923). Zur Kenntnis der Blutkatalase. *Biochem. Z.*, **140**, 63.

United Nations (1968). *International action to avert the impending protein crisis*. UN, New York.

United States President's Science Advisory Committee (1967). *The world food problem*. Govt. Printing Off., Washington.

Van Sumere, C. F., Albrecht, J., Dedonder, A., de Pooter, H. & Pé, I. (1975). Plant proteins and phenolics. In *The chemistry and biochemistry of plant proteins*, ed. J. Harborne & C. F. Van Sumere, p. 211. Academic Press, London.

Varshney, C. K. & Rzoska, J. eds. (1976). *Aquatic weeds in South East Asia*. Junk, The Hague.

Vartha, E. W. & Allison, R. M. (1973). Extractable protein from 'Grasslands Tama' Westerwolds ryegrass. *N.Z. J. exp. Agric.*, **1**, 239.

Vartha, E. W., Fletcher, L. R. & Allison, R. M. (1973). Protein-extracted herbage for sheep feeding. *N.Z. J. exp. Agric.*, **1**, 171.

Venkataswamy, G., Krishnamurthy, K. A., Chandra, P. & Pirie, A. (1976). A nutrition rehabilitation centre for children with xerophthalmia. *Lancet*, **i**, 1120.

Vickery, H. B. (1945). The proteins of plants. *Physiol. Rev.*, **25**, 347.

Vickery, H. B. (1956). Thomas Burr Osborne. *J. Nutr.*, **59**, 3.

Vithayathil, P. J. & Murthy, G. S. (1972). New reactions of *o*-benzoquinone at the thioether group of methionine. *Nature New Biol.*, **236**, 101.

Wallace, G. M. ed. (1975). *Leaf protein concentrates (New Zealand) scene*. Publ. Ruakura agric. Res. Centre, Ruakura, N.Z.

Walsh, K. A. & Hauge, S. M. (1953). Carotene: factors affecting destruction in alfalfa. *J. agric. Fd Chem.*, **1**, 1001.

Wang, J. C. & Kinsella, J. E. (1976a). Functional properties of novel proteins: alfalfa leaf protein. *J. Fd Sci.*, **41**, 286.

Wang, J. C. & Kinsella, J. E. (1976b). Functional properties of alfalfa leaf protein: foaming. *J. Fd Sci.*, **41**, 488.

Wareing, P. F. & Allen, E. J. (1977). Physiological aspects of crop choice. *Phil. Trans. R. Soc., B*, **281**, 107.

Waterlow, J. C. (1962). The absorption and retention of nitrogen from leaf protein by infants recovering from malnutrition. *Br. J. Nutr.*, **16**, 531.

White, J. R., Weil, L., Naghski, J., Della Monica, E. S. & Willaman, J. J. (1948). Protoplasts from plant materials. *Ind. engng Chem.*, **40**, 293.

Widdowson, F. V. (1974). Results from experiments measuring the residues of nitrogen fertilizer given by sugar beet, and of ploughed-in sugar beet tops, on the yield of following barley. *J. agric. Sci. Camb.*, **83**, 415.

Wilkins, R. J. ed. (1977). *Green crop fractionation*. Occasional Symposia 9, Br. Grassld Soc., Maidenhead.

Wilkins, R. J., Heath, S. B., Roberts, W. P. & Foxell, P. R. (1977). A theoretical economic analysis of systems of green crop fractionation. In *Green crop fractionation*, ed. R. J. Wilkins, p. 131. Occasional Symposia 9, Br. Grassld Soc., Maidenhead.

Willaman, J. J. & Eskew, R. K. (1948). *Preparation and use of leaf meals from vegetable wastes. USDA Tech. Bull.*, **958**.

Wilsdon, G. H. C. (1977). The fractionation process at Dengie Crop Driers. In *Green crop fractionation*, ed. R. J. Wilkins, p. 155. Occasional Symposia 9, Br. Grassld Soc., Maidenhead.

Wilson, R. F. & Tilley, J. M. A. (1965). Amino acid composition of lucerne and of lucerne and grass protein preparations. *J. Sci. Fd Agric.*, **16**, 173.

Winterstein, E. (1901). Ueber die stickstoffhaltigen Bestandtheile grüner Blätter. *Ber. dt. bot. Ges.*, **19**, 326.

Withers, N. J. (1973). Production of kenaf under temperate conditions. *N. Z. J. exp. Agric.*, **1**, 253.

Wong, E. (1973). Plant phenolics. In *Chemistry and biochemistry of herbage*, ed. G. W. Butler & R. W. Bailey, vol. 1, p. 265. Academic Press, London.

Woodham, A. A. (1971). The use of animal tests for the evaluation of leaf protein concentrates. In *Leaf protein: its agronomy, preparation, quality and use*, ed. N. W. Pirie, p. 115. Blackwell, Oxford.

Wooten, J. W. & Dodd, J. D. (1976). Growth of water hyacinths in treated sewage effluent. *Econ. Bot.*, **30**, 29.

Worgan, J. T. & Wilkins, R. J. (1977). The utilization of deproteinised forage juice. In *Green crop fractionation*, ed. R. J. Wilkins, p. 119. Occasional Symposia 9, Br. Grassld Soc., Maidenhead.

WHO/U.S.AID (1976). *Vitamin A deficiency and xerophthalmia. Tech. Rep. Ser.*, **590**. WHO, Geneva.

Yamashita, M., Arai, S. & Fujimaki, M. (1976). Plastein reaction for food protein improvement. *J. agric. Fd Chem.*, **24**, 1100.

Yang, S. F. (1970). Sulfoxide formation from methionine or its sulfide analogs during aerobic oxidation of sulfite. *Biochemistry*, **9**, 5008.

Yemm, E. W. (1937). Respiration of barley plants. 3. Protein catabolism of starving leaves. *Proc. R. Soc., B*, **123**, 243.

Yemm, E. W. & Folkes, B. F. (1953). The amino acids of cytoplasmic and chloroplastic proteins of barley. *Biochem. J.*, **55**, 700.

Young, H. E. (1976). Muka: a good Russian idea. *J. For.*, **74**, 160.

Zscheile, F. P. (1973). Long-term preservation of carotene in alfalfa meal. *J. agric. Fd Chem.*, **21**, 1117.

Zubrilin, A. A. (1963) In discussion in Pirie (1963).

Author index

Aceto, N. C., 53
Adadevoh, B. K., 107
Adron, J. W., 81
Agatov, P., 4
Ahmad, B., 34
Ahmad, S. U., 120
Akeson, W. R., 72, 76, 116, 119
Akinrele, I. A., 102
Albrecht, J., 84
Alden, T., 36
Alfin-Slater, R. B., 80, 90
Algeo, J. W., 120
Allen, E. J., 44
Allen, G., vii
Allen, R. J., 113
Allison, F. E., 119
Allison, R. M., 27, 48, 85, 89, 117
Anderson, J. W., 57
Anonymous, 29, 103, 130
Arai, S., 73
Arkcoll, D. B., 21, 23, 25, 28, 48, 53,
    69, 70, 71, 93, 94, 118, 119
Ascarelli, I., 94

Bagchi, D. K., 29, 30, 34, 35, 39, 41,
    42
Bailey, R. W., 66
Balakrishnan, T., 22, 30, 31, 34, 37,
    41
Balasundaram, C. S., 22, 30, 31, 34,
    37, 41, 102
Banyon, E. A., 40
Barber, R. S., 82
Barnes, M. F., 120
Bassir, O., 43
Bates, R. P., 38, 39
Batley, B., 120
Batra, U. R., 20, 28, 29, 42, 48
Bawden, F. C., 6, 23, 65

Beccari, G. B., 1
Begum, A., 93
Belozersky, A., 4
Ben Aziz, A., 94
Berman, T., 119
Berzelius, J. J., 1
Betschart, A. A., 73, 77, 80, 103
Bevenue, A., 11
Bhatia, I. S., 40
Bickoff, E. M., 11, 47, 50, 64, 65,
    66, 71, 76, 77, 86, 93, 94, 96, 103,
    116, 117
Birkinshaw, J. H., 9, 70
Biwshich, N., 4
Blair, A., 81
Blombäck, B., 36
Blunt, C. G., 37
Boerhaave, H., 2, 141
Boncheva, I., 82
Bondi, A., 94
Booth, A. N., 47, 76, 77, 86, 96
Bosshard, H., 87
Boulet, M., 73
Boussingault, J. B., 3
Boyce, D. S., 129, 133
Boyd, C. E., 39, 40
Boyle, J. E., 5
Branscombe, D. J., 66
Braude, R., 82, 117, 132
Bray, W., 55, 67, 73
Bris, E. J., 120
Brisson, G. J., 73
Brookes, I. M., 116, 132
Brown, C. L., 36
Brown, H. E., 42
Bruhn, H. D., 127, 131
Bryant, M. A., 4, 58
Buchanan, A. R., 52, 54, 71, 73, 77,
    78, 79

Budowski, P., 94
Butt, A. M., 120
Byers, M., 21, 23, 27, 28, 30, 31, 33,
   34, 35, 39, 40, 52, 57, 58, 59, 60, 61,
   63, 65, 77, 78, 89, 90, 100, 103,
   109, 115, 121

Camagro, E., 120
Carlsson, R., 21, 28, 40, 41, 42, 45,
   63, 76, 77, 97
Carpenter, K. J., 14
Carr, J. R., 48, 82, 83, 89
Carruthers, I. B., 21, 32, 65
Carter, E. S., 130
Casselman, T. W., 113
Chalmers, M. I., 115
Chan, P. H., 67
Chanda, S., 30, 34, 42
Chandra, P., 93
Chandramani, R., 22, 30, 31, 34, 37,
   41
Chayen, I. H., 12, 65
Cheeke, P. R., 82, 87
Chen, I., 75
Chibnall, A. C., 4, 9, 15, 22, 66
Christ, J. W., 91
Christensen, K. D., 84, 87
Chubinidze, A. S., 132
Clare, N. T., 53
Clark, J. P., 93
Clarke, E. M., 7
Cohen, M., 57
Colker, D. A., 53
Connell, J., 50, 69, 116, 117, 132
Connor, M. A., 47, 77
Corbet, S. A., 38
Cowey, C. B., 81
Cowlishaw, S. J., 72
Crook, E. M., 4, 6, 21, 37, 40

Daniel, V. A., 106
Darwin, C., 48
Davies, M., 66
Davies, R., 86
Davies, W. L., 4
Davys, M. N. G., 12, 13, 18, 19, 20,
   41, 135
Dawson, J. L., 14, 113

de Fremery, D., 50, 64, 65, 66, 71,
   76, 82, 86, 96, 103, 116, 117
de Pooter, H., 84
Dedonder, A., 167
Della Monica, E. S., 5
Derban, L. K. A., 38
Derbyshire, J. C., 113, 114
Deshmukh, M. G., 20, 29, 34, 40,
   42, 48
Dev, D. V., 28, 29, 40, 132
Devadas, R. P., 102, 108
Devi, A. V., 40
Dimaunahan, L. B., 40
Dodd, J. D., 39
Doraiswamy, T. R., 106
Dorner, R. W., 57
Dovrat, A., 119
Duckworth, J., 14, 71, 82
Dumont, A. G., 129, 133
Durand, S., 40
Dutrow, G. F., 36
Dye, M., 91
Dykyjova, D., 40

Edwards, R. H., 64, 65, 76, 86, 94,
   96, 116, 117
Edwardson, W., 130
Eggum, B. O., 84, 87
Eichenberger, W., 54
Ellinger, G. M., 7, 14
Elliott, K., 54
Energy Working Party, Joint
   Consultative Organization for
   Research and Development in
   Agriculture & Food, 130
Ereky, K., 8, 113
Erkomaishvili, S. K., 132
Eskew, R. K., 34, 53
Evans, H. J., 29
Evans, W. C., 66
Eyles, D. E., 72

Fafunso, M., 43, 57, 58, 61, 90
Fairclough, D., 12
Fales, H. M., 53
FAO, 34, 89, 92, 104, 105, 127
Feeny, P. P., 41, 84
Feltwell, J. S. E., 91

Ferrando, R., 96
Festenstein, G. N., 4, 21, 23, 25, 28, 55, 56, 96, 118, 121
Finot, P. A., 51
Fletcher, L. R., 117
Folkes, B. F., 66
Foot, A. S., 71, 82
Ford, J. E., 85, 89
Ford Foundation, 13
Foreman, F. W., 4
Fourcroy, A. F., 1, 3
Fowden, L., 4, 58
Foxell, P. R., 50, 129
Franzen, K. L., 74
Freeman, P. A., 127
Fujimaki, M., 73

Garcha, J. S., 40, 95, 102
Garman, G. R., 87
Gastineau, C., 64, 127
Georgieva, L., 82
Gerloff, E. D., 58, 61
Ghosh, J. J., 35, 39
Gilbert, J. H., 3
Ginoza, W., 57
Gjøen, A. U., 88, 89
Glencross, R. G., 56, 96
Gonzalez, O. N., 40
Goodall, C., 8
Goodall, M., 14, 32, 113
Gordon, C. H., 113, 114
Gore, S. B., 29, 30, 34, 40, 42, 115
Grant, W. F., 83
Green, J., 38
Green, S. H., 100
Green, T. R., 75
Green, V. E., 113
Greenhalgh, J. F. D., 116, 117
Grob, E. C., 54
Grossman, S., 94
Guha, B. C., 11
Guzman, M. A., 35

Hageman, R. H., 57
Hampp, R., 56
Harris, C. E., 117
Hartley, H., 136
Hartman, G. H., 72, 116, 119

Hauge, S. M., 53, 93
Heath, S. B., 21, 27, 129
Heidelberger, M., 65
Henry, K. M., 85, 89
Hentges, J. F., 38
Hepburn, W. R., 82
Herndon, J. H., 53
Herrick, A. M., 36
Hiblits, A. G., 120
Hinde, R., 5
Holden, M., 4, 6, 21, 37, 40, 47, 48, 52, 53, 93, 94, 118
Holdren, R. D., 113, 114
Holl, W., 56
Hollo, J., 119, 128
Holmes, J. G. H., 37
Horigome, T., 77
Horn, M. J., 88
Houseman, R. A., 69, 82, 116, 117, 132
Hove, E. L., 48, 66, 89
Huang, K. H., 73
Hudson, B. J. F., 54, 71, 94
Hudson, J. P., 44
Hudson, W. R., 57
Hume, E. M., 91, 93, 104
Humphries, C., 73
Hussain, A., 34

ICI, 9, 70
Ineritei, M. S., 73
Iovchev, N., 82
Ivins, J. D., 33

Jadhav, B., 34
Jenkins, G., 28
Jennings, A. C., 7, 57
Jones, A. S., 82, 117, 132
Jones, R. J., 37
Jones, W. T., 64
Jönsson, A. G., 120
Joshi, A. L., 132
Joshi, R. N., 20, 28, 29, 30, 34, 40, 42, 43, 48, 73, 111, 115, 117
Julien, J. P., 73

Kamalanathan, G., 102, 108
Kandatsu, M., 77

Kanev, S., 82
Kannangara, C. G., 54
Karis, I. G., 54
Karuppiah, P., 108
Kawatra, B. L., 40, 95, 102
Kempton, T. J., 115
Kendall, F. E., 65
Kennick, W. H., 82
Kiesel, A., 4
King, H. G. C., 56, 96
King, M. W., 27
Kinsburgskii, Z., 132
Kinsella, J. E., 73, 74, 77, 80, 103
Kinzell, J. H., 82
Kitiashvili, D. G., 132
Kleczkowski, A., 23
Knight, J., 54
Knoop, F., 87
Knowles, R., 40
Knuckles, B. E., 64, 65, 66, 76, 86, 96, 116, 117
Koch, L., 119, 121, 128
Koegel, R. G., 127, 131
Kohler, G. O., 47, 50, 64, 65, 66, 71, 76, 77, 82, 86, 93, 94, 96, 103, 116, 117
Kotake, Y., 87
Krebs, H. A., 93, 104
Krishnamoorthy, C. K., 30, 31, 34, 37, 41
Krishnamurthy, K. A., 93
Kunachowicz, H., 88

Laird, W. M., 7, 85, 86
Lala, V. R., 93
Lan, O. B., 38
Lawes, J. B., 3–4, 18, 112
Lazar, M. E., 94
Lea, C. H., 70
Leavenworth, C. S., 4
Leavitt, R. A., 122
Leng, R. A., 115
Levin, D. A., 84
Lewis, C., 130
Lexander, K., 28, 40, 63, 76, 77
Lichtenstein, H., 88
Lamia, I. H., 54, 58, 61
Little, E. C. S., 38

Lohrey, E., 48, 89
Lugg, J. W. H., 4, 65, 66
Lundborg, T., 28, 40, 63, 76, 77

McClure, J. W., 122
McGowan, J. C., 66
McLeod, M. N., 88
Magoon, M. L., 117
Maguire, M. F., 116, 132
Mahadeviah, S., 34, 72, 85
Mahmud, B. A., 30
Mangan, J. L., 64
Marchaim, U., 119
Martin, C., 108
Matai, S., 29, 30, 34, 35, 39, 40, 41, 42
Mathismoen, P., 64
Mauron, J., 51, 80, 88
Mbadiwe, E. I., 7
Mead, J. F., 80, 90
Medlock, O. C., 91
Mendez, J., 35
Menear, J. R., 113
Mfon, B. P., 86
Milic, B. I., 57, 83, 87
Miller, D. S., 89, 95
Miller, R. E., 64, 65, 76, 86, 94, 96, 103, 116, 117
Mitchell, D. S., 38
Mitchell, H. L., 75
Mitchell, J. C., 88
Mitchell, K. G., 82
Mitscher, L. A., 122
Mokody, S., 73
Monsod, G. G., 39
Moore, P. D., 119
Morozova, A., 132
Morris, T. R., 129
Morrison, J. E., 32, 48, 83, 99
Mottola, A. C., 93
Mummery, R., 91, 92
Mungikar, A. M., 29, 30, 34, 42, 115, 117
Munshi, S. K., 72
Murthy, G. S., 87
Myer, R. O., 82

Naghski, J., 5
Narayan, J., 28

Nasir, M., 40
Nass, M. M. K., 5
National Academy of Sciences, USA, 39
National Economic Development Office, 114
Naumenko, V., 132
Nedorizescu, M., 37
Nehru, J., 28
Njaa, L. R., 88, 89
Nolan, J. V., 115
Nowakowski, T. Z., 121
Nutrition Society, 47

Oelshlegel, F. J., 34
Oke, O. L., 86, 102, 107
Olatunbosun, D. A., 107
Omole, T. A., 86
Orlandi, W., 120
Ornes, W. H., 39
Osborne, T. B., 4, 9, 61
Overbury, T., 33
Oyakawa, N., 120

Paredes-Lopez, O., 120
Parr, W. H., 66, 70
Patil, R. E., 132
Patriquin, D. G., 40
Pawlowa, M., 4
Pé, L., 84
Pearson, G., 48, 82, 83, 89
Pereira, S. M., 93
Petersen, D. W., 51
Pienazek, D., 88
Pierpoint, W. S., 65, 84, 87
Pieterse, A. H., 39
Pirie, A., 92, 93
Pirie, N. W., vii, 2, 5, 6, 10, 11, 12, 13, 17, 18, 19, 20, 21, 28, 32, 34, 40, 41, 42, 44, 45, 47, 48, 51, 55, 57, 63, 64, 65, 67, 70, 72, 83, 88, 91, 95, 99, 100, 102, 103, 108, 113, 114, 120, 126, 128, 134, 135
Pleshkov, B. P., 58
Pope, J. A., 81
Powling, W. T., 14
Prasad, V. L., 132

Prasanappa, G., 69
Protein Advisory Group, 127
Proust, J. L., 138
Pusztai, A., 57

Rakowska, M., 88
Ramadoss, C., 31, 34, 37, 41
Ramana, K. V. R., 40, 57, 61, 85, 86, 89, 94
Rancour, J. M., 75
Randolph, J. W., 113
Rangnekar, D. V., 132
Rao, B. S., 93
Rao, C. N., 93
Rao, N. A. N., 40
Raymond, W. F., 34, 72, 117
Ream, H. W., 119
Reddy, V., 86, 93
Rees, M. W., 66
Reid, G. W., 116, 117
Rey, J., 2
Riaz-ud-Din, 73, 77, 79
Richardson, M., 75
Richardson, T., 54, 75
Riel, R. R., 73
Ries, S. K., 122
Rivera-Brenes, L., 113
Robbins, R. C., 39
Roberts, E. A. H., 87, 129
Robson, T. O., 38
Rodriguez, E., 88
Roja, F. C., 122
Rothschild, M., 91, 92
Rouelle, G. F., 1–2
Rouelle, H. M., 2–3, 63, 79, 138–41
Rowan, K. S., 57
Roy Chowdhury, S., 35, 39
Ryan, C. A., 75
Rzoska, J., 38

Sakano, K., 67
Salam, A., 73, 77, 79
Saldana, G., 42
Sale, P. J. M., 33
Samuel, D. M., 102
Samuel, P. D., 89
Saunders, R. M., 47, 76, 77, 86, 96

Savangikar, V. A., 29, 43, 73
Savel'ev, D. N. O., 54
Schalen, V., 28, 40, 63, 76, 77
Scherp, H. W., 65
Schnabel, C. F., 9
Schroeder, J. R., 34
Schubert, K. R., 29
Schwartz, K., 55
Scrimshaw, N. S., vii, 35
Sentheshanmuganathan, S., 40
Shah, F. H., 30, 40, 48, 73, 77, 103, 120
Shardlow, A. W., 12
Shumway, L. K., 75
Shurpalekar, K. S., 85, 89
Simonsson, A., 28, 40, 63, 76, 77
Singh, N., 17, 20, 31, 34, 40, 41, 47,
    48, 57, 61, 69, 72, 75, 85, 86, 89, 94,
    95, 106
Singh, S., 67
Siren, G., 36
Skinner, F. A., 122
Slade, R. E., 9, 15, 66, 70
Slansky, F., 41
Slesser, M., 130
Smith, D., 119
Smith, D. A., 11
Smith, D. C., 5
Smith, R. A., 122
Smith, R. C., 89
Smith, R. S., 12, 65
Somaroo, B. H., 83
Spais, A., 96
Spencer, R. R., 93
Squibb, R. L., 35
Stahmann, M. A., 34, 54, 58, 61, 72,
    76, 116, 119
Staron, T., 74, 77
Stein, E. R., 42
Steinberg, D., 53
Stojanovic, S., 87
Straub, R. J., 131
Street, G., 18, 19, 20
Stumpf, P. K., 54
Sturrock, J. W., 21, 23, 27, 33, 115
Subba Rao, M. S., 69
Subba Rau, B. H., 40, 47, 57, 61, 72,
    85, 86, 89, 95
Sullivan, J. T., 11

Sundaravalli, O. E., 85
Sur, B. K., 95
Sutton, D. L., 39
Sweely, C. C., 122
Synge, R. L. M., 7, 57, 84, 85, 86, 115

Tao, M. C., 73
Tapper, B. A., 48
Tarasenko, A., 132
Taylor, D. L., 5
Taylor, K. G., 39
Tekale, N. S., 30, 34, 42
Thakur, M. L., 83
Thapar, V. K., 72
Thirkell, D., 12, 65
Thomas, F. H., 113
Thresh, J. M., 57
Thung, T. H., 57
Tilley, J. M. A., 65, 72
Toosy, R. Z., 103
Topps, J. H., 89
Towers, G. H. N., 88
Tracey, M. V., 4, 6, 20, 48
Tragardh, C., 66
Trench, R. K., 5
Trigg, T. E., 89
Tristram, G. R., 12, 65
Tsuchihashi, M., 65

Uhlendorf, B. W., 53
Ullah, M., 34
United Nations, 127
United States Presidential
    Advisory Committee, 127
Urs, M. K., 89

Valadon, L. R. G., 91, 92
Valente, E. O. L., 120
Van Sumere, C. F., 84
Varshney, C. K., 38
Vartha, E. W., 27, 117
Vauquelin, L. N., 1, 3
Venkataswamy, G., 93
Vickery, H. B., 4
Vijayaraghavan, P. K., 40
Vithayathil, P. J., 87

Wagle, D. S., 40, 72, 95, 102
Wakeman, A. J., 4
Walgenbach, R. P., 119
Wallace, G. M., 22, 66, 117
Walsh, K. A., 53, 93
Wang, J. C., 103
Want, J. P. H. van der, 57
Wareing, P. F., 44
Warwick, M. J., 71, 94
Waterlow, J. C., 105
Watt, W. B., 7, 57
Watts, E., 38
Waygood, E. R., 57
Webb, T., 12, 65
Weil, L., 5
Wert, V., 122
White, J. R., 5
WHO/U.S. AID, 91, 92
Widdowson, F. V., 32
Wildman, S. G., 57, 67
Wilkins, R. J., 69, 121, 129
Willaman, J. J., 5, 34
Williams, K. T., 11

Wilsdon, G. H. C., 127
Wilson, R. F., 65
Winfree, J. P., 113
Winterstein, E., 4, 11
Withers, N. J., 35
Womack, M., 88
Wong, E., 57
Wooden, G. R., 120
Woodham, A. A., 71, 81, 82, 83, 86
Woodruff, M. F. A., 11
Wooten, J. W., 39
Worgan, J. T., 121

Yamashita, M., 73
Yang, S. F., 88
Yemm, E. W., 4, 66
Young, H. E., 36

Zia-ur-Rehman, 30
Zimmermann, C., 73
Zscheile, F. P., 53
Zubrilin, A. A., 11

Subject index

Abbreviation used: LP for leaf protein

acid: facilitates filtration of LP, 51; increases loss of carotene, 51; removes magnesium from chlorophyll, 51, 52; for preservation of moist LP, 69, 133

Agave sisalana (sisal), 35

age of leaves, and yield of LP, 4, 21, 26

alkali: increases extraction of LP from some leaves, 39, 47; increases extraction of pectic substances, 47, 55; precipitates free alkaloids, 122; stabilises carotene, 47, 93; treatment of straw with, 128

alkaloids, 44; removed from LP by washing at pH 4, 56, 96, and from 'whey' by precipitation with alkali, or solvent extraction, 122

Allmania modiflora, 40

Amaranthus caudatus, 76, 108

amino acid composition of LP, 15, 57–61, 95; compared with that suggested by FAO for protein suitable for children, 104–5

amino acids, free, in 'whey', 120

animal feed: 'chloroplast' protein as, 64; fibre as, 9, 115–16; nutritive value of LP for, 80–3; unfractionated leaf juice as, 14, 132; ways of using LP as, 131–3; 'whey' in, 130

annual plants *v.* perennials, as sources of LP, 27

antiproteases (thermostable), in leaves, 75

Arachis hypogaea (groundnut), 34

ascorbic acid, in lucerne juice, 122

ash content of LP, 55

Aspergillus, 'whey' as medium for, 121

Atriplex caudatus, *A. hortensis*, saponins in, 97

Atriplex hortensis, *A. latifolium*, different fractions from, 63

autolysis, in leaf juice, 14, 20, 48

Avena sativa (oats), as leafy vegetable, 42

bacteria, 'whey' as medium for, 121

bamboo, 36

banana and LP pie, 101–2

Basella alba, 43

beans, silage from fibre of, 117; *see also Phaseolus*, *Vicia*

Beta vulgaris (sugar beet): LP from tops of, 14, 32; silage from fibre of, 117

biochanin A, estrogenic substance in clover, 96

Boehmeria nivea (ramie), 35

Bracharia mutica (grass), 30

Brassica carinata (oilseed), 42

Brassica napus: rape, 28, 63; turnip, 42

Brassica nigra (oilseed), 42

Brassica oleracea: cabbage, 4; kale, 21

by-product leaves, as source of LP, 14, 31–6

Candida utilis, 'whey' as medium for, 121

carbohydrates: in LP, 55; in 'whey', 55, 118, 119, 120–1

β-carotene (pro-vitamin A): in 'chloroplast' fraction, 63; consumption of vegetables containing, and serum retinol of

β-carotene (pro-vitamin A)—*cont.*
children, 93; in LP, 53–4, 72, 94, 95;
loss of, during processing, 94, 118;
tropical vegetables rich in, 92
carotenoids, in LP, 53–4; acid
increases loss of, 51; alkali
increases stability of, 47, 93
Celosia argentea, as source of LP for
malnourished children in
Nigeria, 108
Cenchrus glaucus (grass), 30
centrifuging, separation of
'chloroplast' fraction by, 64
cereals: flavour of LP from, 98;
supplementation with LP of diets
based on, 94–5
cheese: addition of LP to, 70;
preservation of moist LP in form of,
9, 70
Chenopodium album (fat hen), 21–2
Chenopodium quinoa (quinoa), 42, 97
children: growth of, with LP and
other supplements, 106–7, 109–11;
LP as protein supplement for,
42, 104-5; LP as supplement to
inadequate supply of milk for,
105–6; recovery of, from
kwashiorkor, with LP supplements,
107–8
Chinese cabbage, 60
chlorophyllase, 48, 52, 83
chlorophyllides, 48, 52
chlorophylls: breakdown products of,
52; structure of, 53
'chloroplast' fraction of LP, 60, 62-3;
digestibility of, 76, 78, 85;
methods of separating, 63–7
chloroplasts: in LP, 5–6, 62–3; in
fibre, 6–7
Chrozophora sp., 40
chutney, containing LP, 103
clovers: flavour of LP from, 98;
estrogens in, 56; *see also Melilotus,
Trifolium*
coagulation (precipitation): of
'cytoplasm' fraction by heat, 64;
of leaf juice, by acid, 47, by heat,
14, 47, 48–50, 63, 75, 132–3,

spontaneously, 14, 132, and by
trichloroacetic acid, 17, 20
commercial fodder fractionation,
127–8
computer analysis, of amino acid
composition of 39 preparations of
LP, 59, 61
cooking methods, for LP, 98, 99,
101
coppicing, of trees for harvesting
leaves, 36
Corchorus olitorus, 43
Corchorus sp. (jute), 35
coumestrol, estrogen in lucerne, 96
crop plants, LP from, 23–30
Crotalaria juncea (sunn hemp), 35
curry cubes, containing LP, 101
Cyamopsis psoralioides (guar), 120
cyst(e)ine: in amino acid analyses of
LP, 61; modification of, in
processing, 87–90
'cytoplasm' fraction of LP, 60,
62–3; digestibility of, 76, 78, 85;
methods of separating, 63–7;
nearly free from lipid, 80

Dactylis glomerata (cocksfoot grass),
25, 27
Daucus carota (carrot), 86, 117
dhal balls, containing LP, 102–3
digestibility of fibre, 116–17
digestibility of LP; affected by
temperature and rate of drying,
71–2; of 'chloroplast' and
'cytoplasm' fractions, 62, 76, 78,
85; decreases with excessive
heating, 78, but is restored by
solvent drying, 72–3; increases on
coagulation by heat, 75, and with
nitrogen content, 77; *in vitro*,
freeze-dried LP, 76–8, and
heat-damaged LP, 78–80; *in vivo*,
modification of cyst(e)ine and
methionine and, 87–90, phenolic
compounds and, 83–7, unsaturated
fatty acids and, 90
dry matter content: of crop,
fertiliser and, 21, 26; of crop,

period between harvests and, 27; of coagulated LP, 51; distribution of, between LP, fibre and 'whey', 112, 114, 118

drying: of crops, 113–14; of fibre from LP production, 114, 117, 135; of LP, in air current, 71–2, and by freeze-drying, 70

dyes (indigo, madder) might be by-products of LP, 122

economic assessment, of fodder fractionation, 128–31

Eichhornia crassipes (water hyacinth), 11, 39, 95, 120

Endomycopsis fibuliger, 'whey' as medium for, 121

energy: areas with diets lacking in, 136; diminishing requirement of, for processing LP, 136; required for removal of moisture from crop by pressing and by drying, 135

enzymes: during heat coagulation 48; in leaf extracts, 6, 15

estrogens, in clovers, 56, 96

eutrophication, and growth of water weeds, 38

extenders for LP (flour, etc.), drying after mixture with, 71, 82

extractability of LP, 21; decreases with aging of leaves, 26, and with fibre content, 30; increases with addition of alkali, 39, 41, with protein content, 21, and with washing of crop, 46; in legumes, 27

fatty acids: age of leaves, and pattern of, 54; unsaturated, reactions of, 51, 79

fertiliser, nitrogenous: and nitrogen content of LP, 76; 'whey' as, 119, 121; and yields of LP, 26, 27

ferulic acid derivative, antioxidant, in lucerne, 94

Festuca arundinalis (tall fescue grass), 21

fibre (fibrous residue), 112–15; for animal feed, 9, 115–17; enzymes associated with, 93; extractability of LP decreases with increasing content of, 30; nitrogen attached to, 6–7, 16, 20, 115–16; pressing of, for silage or drying, 5, 135; processes disintegrating, 16; silage from, 117–18

filtration: fractionation by, 66–7; of LP suspended in water, facilitated by addition of salt or acid, 51

flavour: of LP from different plants, 44, 98; of moist LP, affected by irradiation, 69; should be diminished as far as possible, 50

foods, addition of LP to, 99, 100–4, 126

formononetin, estrogen in clover, 96

fractionation of LP, into 'chloroplast' and 'cytoplasm' parts, 61–3; methods for, 63–7

Fragaria vesca (strawberry), 21

froth: intractable amounts of, prevent extraction of LP from some species, 22

fungi: plants heavily attacked by, not suitable for LP production, 56; 'whey' as medium for, 121

gibberellic acid applied to crop, and yield of LP, 28

Gliricidia, 37, 102

Glycine max (soybean), 73

Gossypium hirsutum (cotton), 35, 102

grasses: extraction of LP from, 22; juice of, as animal feed, 132, and as medicine, 9; nitrogen content of crop, and of fibre from, 116; position of protein in leaves of, 27; silica in LP from, 55

green manuring, by-products from LP extraction for, 30

growth depressants, in 'whey' from lucerne, 51, 96

growth regulator (Simazine) applied to crop, and yield of LP, 29

growth stimulants in 'whey' ?, 121–2

harvesting: frequency of, on yield of LP, 25, 30; type of machine used for, and extractability of LP, 46

heating: coagulation by, of 'cytoplasm' fractions, 63–4, and of leaf juice, 47, 48–50, 63, 75, 132–3; excessive, decreases digestibility, 78; in pulping machine, 10

Helianthus annuus (sunflower), 28, 63

herbicides, should not be used on crops for LP, 46

Hibiscus cannabinus (kenaf), 35

Hibiscus sabdariffa (roselle), 35

Hordeum vulgare (barley), 24, 60

hydroxyproline, in fibre, 7

infant-feeding, LP for, 42, 115–16; *see also* children

insecticides, should not be used on crops for LP, 46

Ipomoea batatas (sweet potato), 22

irradiation, flavour of moist LP affected by, 69

Justicia americana (water weed), 40

lactones: react with cysteine, may cause dermatitis, 87–8

lead, as impurity in LP, 56

leaf juice: discarding of different fractions of, by different workers, 124; fibre fragments in, 46–7; little obtained by pressing only, 113; nonprotein nitrogen in, 27; preparation of LP from, *see under* leaf protein; starch in, 55; technique of extraction of, 16–20; unfractionated, as animal feed, 14; washing of material for extraction of, 46

leaf protein (LP): arguments for use of, 15–16, 124; composition of, 51–61; digestibility of, *see* digestibility; division of, into 'chloroplast' and 'cytoplasm' fractions, 63; early work on, 2–4, (Rouelle's paper),

138–41; extractability of, *see* extractability; fractionation of, 61–7; prelude to production of, 8–14; preparation of unfractionated, 47–51; proposal to compress, and mature, as 'cheese', 9, 70; selection of plant species for, 21–3, (by-product leaves), 31–6, (crop plants), 23–30, (miscellaneous) 40–5, (tree leaves) 36–7, (water weeds), 37–41; yield of, per hectare, 24, 25, 27, 126

leaves: rate of deterioration of, after harvest, 20; type of, suitable for extracting LP, 21, 23

legumes, 28–9, 30; antiproteases in leaves of, 75; *see also individual species*

Leonotis taraxifolia, unacceptable flavour of LP from, 44

Leucaena glauca, 37

light, affects LP in storage, 53, 79–80

lipids, in LP, 52–4, 73; changes in, during heating of LP, 79–80; solvent drying and, 71, 72; toxicity of peroxidised, 80, *see also* fatty acids

Lolium multiflorum (ryegrass), 27

Lolium perenne (ryegrass), 27, 116, 118

Lupinus albus (lupin), 60, 61, 63, 89

lysine: complexes of phenolics with, 85; Maillard reaction between reducing sugars and, 51, 69

machinery for extraction and processing of LP: batch extractor, 13; belt press, 135–6; for coagulation of leaf juice, 49; Ereky's and Goodall's, 8; hammer-mill pulper, 12; IBP pulper and press, 18–19, 20; improvements needed in, 133–7; oil expeller, 14; tests of different types, 10

Maillard (browning) reaction between lysine and reducing sugars, 51, 69

Manihot esculenta (*utilissima*) (cassava), 31, 108

manufacture of LP: commercial, 127–8; economic assessment of, 127–31; small-scale local, 74, 125, 126–7

Medicago sativa (lucerne), 4, 7; components of, (antioxidant) 94, (chlorophyllase) 48, (coumestrol) 96, (growth depressants) 51, 96, (saponins) 56; froth in processing some varieties of, 22; leaf juice from, as animal feed, 132; as leafy vegetable, 42; LP from, (flavour) 44, 98, (fractionation) 63, (sulfur content) 89, (yield) 27, 28–9; nitrogen contents of crop and fibre, 116; silage from fibre, 117; 'whey' from, 120

Melilotus spp., 27

methionine, in LP, 60, 61; conversion of, to sulfoxide and sulfone, 88–90; modification of, in processing, 87–8

microorganisms: avoidance of contamination of LP with, by dairying techniques, 70; decomposition of leaf juice by, 68; in moist LP preserved with acid, 69; 'whey' as medium for, 66, 112, 120

mimosine, toxic amino acid in *Leucaena*, 37

mitochondria, in extraction of LP, 6

Mucor racemosus, usual contaminant of preserved moist LP, 69–70

Musa sapientum (banana), 36; *M. textilis* (abaca), 36

mycotoxins, 56

Nicotiana, crystals containing ribulose-diphosphate carboxylase from, 67

nitrogen: content of, in orginal crop and in extracted fibre, 6–7, 115–16, and in 'whey', 118; nonprotein, in leaf juice, 27; ratio of nonprotein to protein, differs between and within species, 22

nucleic acids: carbohydrate from, 55;

delay between pulping and coagulation favours removal of, 49; precipitated by trichloracetic acid, 17; in ribosomes, 6

nutritive value of LP: affected by methods of preparation more than by species differences, 81, 90, 94; in diets for children, *see under* children; in diets for pigs, 82–3, 89; effects on, of modification of cyst(e)ine and methionine, 87–90, and of phenolic and other tanning agents, 83–7

Nymphaea lotus, *N. odorata* (water lilies), 40

oxalate: in sugar beet tops, 32; in vegetable leaves, 42

oxidation: fractionation in presence of inhibitors of, 64; of LP, enzymic and nonenzymic, 48, 70; of unsaturated fatty acids, increased by acid, 51

palatability, tests 10, 103–4, 120,

Panicum maximum (grass), 30

papain, digestibility of LP by, 77

Parthenium hysterophorus, troublesome weed in India, 43

patents, 3, 8, 9, 12, 64, 113

pectic substances, alkali increases extraction of, 47, 55

pectin methyl esterase, and pH during extraction, 47

penicillin, 'whey' as growth medium for production of, 120

Pennisetum typhoideum × *P. purpurea* (hybrid napier grass), 29, 30, 115, 117

pepsin, followed by trypsin: digestibility of LP by, 76–7

pH: during extraction of LP, 5; during fractionation, and protein content in 'chloroplast' fraction, 63; of leaf juice, 47; at which LP coagulates, 47

Phaseolus aureus, tetraploid form (tetrakalai), 29

Phaseolus mungo (mung bean), 120
Phaseolus vulgaris (runner bean), 4
phenolic substances in leaf juice,
    56–7; combine with LP if contact
    of coagulum with 'whey' is
    prolonged, 50; and digestibility,
    83–6, (increase requirement for
    methylating agents?) 86, 87;
    diminish extraction of LP, 23, 36;
    oxidation of, to tanning agents, 6,
    86, (inhibited by alkaline sulfite) 47;
    react with methionine and cysteine?
    87–8
pheophorbides (photosensitisers),
    risk of producing, 48, 51, 52, 83
pheophytins, 51, 52
phosphate, alkaline: fractionation by
    addition of, 67
phosphorus content, of 'whey', 119
photosensitisation, by pheophorbides,
    48, 83
Phragmites (reed), 40
phylloerythrin, 52, 53
phytotoxicity, of 'whey', 119
pigs: leaf juice as protein food for,
    132; nutritive value of LP in
    diets for, 82–3, 89
Pistia stratiotes (Nile cabbage), 39
Pisum sativum (pea), 33, 98, 116, 120
plastoquinone, in chloroplasts, 85
polyvinylpyrrolidone, for sequestration
    of tanning agents, 57
potassium, in 'whey', 119
preservation: of leaf juice, 68–9; of
    moist LP, 69–70
pressing: of coagulated LP, 50; in
    extraction of LP, 5; for silage, 135
proteases, in leaf juice, 68, 75
protein: history of recognition of,
    1–2; in leaves, 2–4; *see also* leaf
    protein
Pueraria phaseoloides (kudzu), 30

Raphanus sativus (rodder radish), 25,
    26, 27
regrowth, of plants used for LP, 24,
    125

retinol, in serum of children receiving
    LP and other food supplements, 93
rhizobia: inoculation of lucerne with,
    and yield of LP, 29; of *Vigna*, 29;
    'whey' as medium for, 120
Rhodotorula sp., 'whey' as medium for,
    120
ribonuclease, in leaf juice, 6, 17
ribosomes, in extraction of LP, 6
1, 5-ribulose-diphosphate carboxylase,
    dominant component of leaf juice,
    6, 58; crystals of, 67; in 'cytoplasm'
    fraction, 64
rice, LP as supplement to diet of, 95
rubbing or grinding, of leaves for
    extraction of LP, 5, 8, 113, 134

Saccharomyces spp., 'whey' as
    medium for, 120
Saccharum officinarum (sugar cane), 31
salting, preservation of moist LP by,
    69
Salvinia auriculata (water fern), 39
Sambucus niger (elder), 37
saponins: in legume seeds, 96–7; in
    lucerne, 56
seaweeds, 40
Secale cereale (rye), 24, 26
seeds, proteins of, 3, 15
Sesbania sesban, 30
silage: from fibre, 117; pressing of
    material for, 135
silica, in LP from grasses, 55
Sinapis alba (mustard), 25, 26, 28, 119
solanin (glycoalkaloid), in potato
    haulm, 32, 51, 122
Solanum incanum, 43
Solanum nodiflorum, 43
Solanum tuberosum (potato):
    antiproteases in leaves of, 75;
    centrifuging of leaf juice from, 65;
    LP from, 21, 98; silage from fibre
    of, 117; 'whey' from, 50–1
soluble material in LP, recommended
    maximum amount of, 51
solvents: drying by extraction with,
    72–3; extraction of alkaloids by,
    122; fractionation by, 65

Spinacia oleracea (spinach), 4
starch, in fibre, 55
stem, contributes little protein, 27
steroids, in agaves and yams, 122;
   see also estrogens
storage of LP, retention of carotene
   during, 53, 94
straw, alkali treatment of, 128
Streptomyces: possible inoculation of
   preserved moist LP with, to
   produce vitamin B 12, 70
sugar, as preservative for moist LP, 69
sulfite, as preservative for leaf juice,
   47, 57, 68
sulfur in LP: in cyst(e)ine and
   methionine only, 61; deficient in
   seed and leaf proteins, 87, 89
Symphytum asperrimum, S. officinale
   (comfrey), 22

Talinum triangulare, 43
tannins: diminish digestibility of LP,
   84–5; diminish extraction of LP, 23
Thalassia testudinum (turtle grass), 40
Tithonia tagetiflora (grown for
   flowers), 42
tocopherol (vitamin E), as natural
   antioxidant in LP, 70–1, 94
tomato, antiproteases in leaves of, 75
tree leaves, LP from, 36–7
Trifolium alexandrinum (berseem),
   29–30, 115, 117, 120
Trifolium pratense (red clover), 20,
   26–7, 96
Trifolium spp., 27
Trigonella foenum-graecum (fenugreek),
   22
Triticum aestivum (wheat), 20, 24;
   LP from, 90, 95; 'whey' from, 119;
   yield of LP from, 25, 26

Tropaeolum majus (nasturtium), 63
tryptophan, in amino acid analyses of
   LP, 61
Typha (reed), 40

Urtica dioica (nettle), 44–5, 63

varieties of one species: differences in
   LP yield among, 21–2
vegetables: dark green leafy, as
   source of vitamin A, 91–3; LP
   from discarded leaves of, 34
Vicia faba (bean), 57, 76
Vicia sativa (tares), 27–8
Vigna sinensis, V. unguiculata
   (cowpea), 29, 72, 98, 108
vitamin A, 63
vitamin B group, 55
vitamin E, 55, 71
vitamin K, 55

washing: of plant material before
   extraction of LP, 46; of
   unfractionated LP, 51, 77
water weeds: LP from, 37–40;
   processing machinery for, on a
   barge? 136; silage from fibre of,
   rejected by cattle, 117
weeds, production of LP from, 43,
   44–5
'whey' from LP coagulum, viii, 50,
   51; quality and uses of, 118–23

xanthophyll, in LP, 53, 127
xerophthalmia, dark green leafy
   vegetables in cure of, 93

yeasts, 'whey' as medium for, 120

Zea mays (maize), 21, 24, 33–4